NOTES

OF A COURSE OF NINE LECTURES ON

LIGHT

DELIVERED AT

THE ROYAL INSTITUTION OF GREAT BRITAIN

APRIL 8—JUNE 3, 1869

BY

JOHN TYNDALL, LL.D. F.R.S.

LONDON:

LONGMANS, GREEN, AND CO.

1870.

PREFACE.

THESE NOTES were prepared for the use of those who attended my Lectures on Light last year, and were not intended for further publication. Enquiries and requests regarding them from teachers and students who have read them, cause me now to think that the Notes may be useful beyond their contemplated limits. The Messrs. Longman have therefore undertaken their publication in a very cheap form.

To my friend Professor Goodeve, who has been kind enough to look over the proofs, my best thanks are due and tendered.

ROYAL INSTITUTION: *May*, 1870.

a

CONTENTS.

———•O•———

	PAGE
General Considerations. Rectilinear Propagation of Light	1
Formation of Images through small Apertures	2
Shadows	,,
Enfeeblement of Light by Distance ; Law of Inverse Squares	3
Photometry, or the Measurement of Light	4
Brightness	5
Light requires Time to pass through Space	6
Aberration of Light	7
Reflexion of Light (Catoptrics)—Plane Mirrors	8
Verification of the Law of Reflexion	,,
Reflexion from Curved Surfaces : Concave Mirrors	11
Caustics by Reflexion (Catacaustics)	13
Convex Mirrors	14
Refraction of Light (Dioptrics)	15
Opacity of Transparent Mixtures	19
Total Reflexion	20
Lenses	22
Converging Lenses	,,
Diverging Lenses	,,
Vision and the Eye	23
Adjustment of the Eye: Use of Spectacles	24
The Punctum Cœcum	25
Persistence of Impressions	26
Bodies seen within the Eye	27
The Stereoscope	28
Nature of Light ; Physical theory of Reflexion and Refraction	29

Contents.

	PAGE
Theory of Emission	30
Theory of Undulation	31
Prisms	34
Prismatic Analysis of Light: Dispersion	"
Invisible Rays: Calorescence and Fluorescence	35
Doctrine of Visual Periods	36
Doctrine of Colours	37
Chromatic Aberration. Achromatism	38
Subjective Colours	39
Spectrum Analysis	40
Further Definition of Radiation and Absorption	41
The pure Spectrum: Fraunhofer's Lines	42
Reciprocity of Radiation and Absorption	43
Solar Chemistry	44
Planetary Chemistry	45
Stellar Chemistry	"
Nebular Chemistry	46
The Red Prominences and Envelope of the Sun	"
The Rainbow	"
Interference of Light	48
Diffraction, or the Inflexion of Light	49
Measurement of the Waves of Light	52
Colours of Thin Plates	54
Double Refraction	57
Phenomena presented by Iceland Spar	59
Polarization of Light	60
Polarization of Light by Reflexion	62
Polarization of Light by Refraction	63
Polarization of Light by Double Refraction	"
Examination of Light transmitted through Iceland Spar	64
Colours of Double-refracting Crystals in Polarized Light	65
Rings surrounding the Axes of Crystals in Polarized Light	68
Elliptic and Circular Polarization	69
Rotatory Polarization	70
CONCLUSION	71

NOTES

ON

L I G H T.

———•◦•———

General Considerations. Rectilinear Propagation of Light.

1. The ancients supposed light to be produced and vision excited by something emitted from the eye. The moderns hold vision to be excited by something that strikes the eye from without. What that something is we shall consider more closely subsequently.

2. *Luminous* bodies are independent sources of light. They generate it and emit it, and do not receive their light from other bodies. The sun, a star, a candle-flame, are examples.

3. *Illuminated* bodies are such as receive the light by which they are seen from luminous bodies. A house, a tree, a man, are examples. Such bodies scatter in all directions the light which they receive; this light reaches the eye, and through its action the illuminated bodies are rendered visible.

4. All illuminated bodies scatter or reflect light, and they are distinguished from each other by the excess or *defect* of light which they send to the eye. A white cloud in a dark-blue firmament is distinguished by its excess of light; a dark pine-tree projected against the same cloud is distinguished through its defect of light.

5. Look at any point of a visible object. The light comes from that point in straight lines to the eye. The lines of light, or *rays* as they are called, that reach the pupil form a *cone*, with the pupil for a base, and with the point for an apex. The point is always seen at the place where the rays which form the surface of this cone intersect each other, or, as we shall learn immediately, where they *seem* to intersect each other.

6. Light, it has just been said, moves in straight lines; you see a luminous object by means of the rays which it sends to the eye, but you cannot see round a corner. A small obstacle that intercepts the view of a visible point is always in the straight line between the eye

B

and the point. In a dark room let a small hole be made in a window-shutter, and let the sun shine through the hole. A narrow luminous beam will mark its course on the dust of the room, and the track of the beam will be perfectly straight.

7. Imagine the aperture to diminish in size until the beam passing through it and marking itself upon the dust of the room shall dwindle to a mere line in thickness. In this condition the beam is what we call a *ray* of light.

Formation of Images through small Apertures.

8. Instead of permitting the *direct sunlight* to enter the room by the small aperture, let the light from some body illuminated by the sun—a tree, a house, a man, for example—be permitted to enter. Let this light be received upon a white screen placed in the dark room. Every visible point of the object sends a straight ray of light through the aperture. The ray carries with it the colour of the point from which it issues, and imprints that colour upon the screen. The sum total of the rays falling thus upon the screen produces an *inverted* image of the object. The image is inverted because the rays *cross* each other at the aperture.

9. *Experimental Illustration.*—Place a lighted candle in a small camera with a small orifice in one of its sides, or a large one covered by tinfoil. Prick the tinfoil with a needle; the inverted image of the flame will immediately appear upon a screen placed to receive it. By approaching the camera to the screen, or the screen to the camera, the size of the image is diminished; by augmenting the distance between them, the size of the image is increased.

10. The boundary of the image is formed by drawing from every point of the outline of the object straight lines through the aperture, and producing these lines until they cut the screen. This could not be the case if the straight lines and the light rays were not coincident.

11. Some bodies have the power of permitting light to pass freely through them; they are *transparent* bodies. Others have the power of rapidly quenching the light that enters them; they are *opaque* bodies. There is no such thing as perfect transparency or perfect opacity. The purest glass and crystal quench some rays; the most opaque metal, if thin enough, permits some rays to pass through it. The redness of the London sun in smoky weather is due to the partial transparency of soot for the red light. Pure water at great depths is blue; it quenches more or less the red rays. Ice when seen in large masses in the glaciers of the Alps is blue also.

Shadows.

12. As a consequence of the rectilinear motion of light, opaque bodies cast shadows. If the source of light be a *point*, the shadow is sharply defined; if the source be a luminous *surface*, the perfect shadow is fringed by an imperfect shadow called a *penumbra*.

13. When light emanates from a point, the shadow of a sphere placed in the light is a *divergent* cone sharply defined.

14. When light emanates from a luminous globe, the perfect shadow of a sphere equal to the globe in size will be a *cylinder*; it will be bordered by a penumbra.

15. If the luminous sphere be the larger of the two, the perfect shadow will be a *convergent cone*; it will be surrounded by a penumbra. This is the character of the shadows cast by the earth and moon in space; for the sun is a sphere larger than either the earth or the moon.

16. To an eye placed in the true conical shadow of the moon, the sun is totally eclipsed; to an eye in the penumbra, the sun appears horned; while to an eye placed beyond the apex of the conical shadow and within the space enclosed by the surface of the cone produced, the eclipse is *annular*. All these eclipses are actually seen from time to time from the earth's surface.

17. The influence of magnitude may be experimentally illustrated by means of a batswing or fishtail flame; or by a flat oil or paraffin flame. Holding an opaque rod between the flame and a white screen, the shadow is sharp when the *edge* of the flame is turned towards the rod. When the broad surface of the flame is pointed to the rod, the real shadow is fringed by a penumbra.

18. As the distance from the screen increases, the penumbra encroaches more and more upon the perfect shadow, and finally obliterates it.

19. It is the angular magnitude of the sun that destroys the sharpness of solar shadows. In sunlight, for example, the shadow of a hair is sensibly washed away at a few inches distance from the surface on which it falls. The electric light, on the contrary, emanating as it does from small carbon points, casts a defined shadow of a hair upon a screen many feet distant.

Enfeeblement of Light by Distance; Law of Inverse Squares.

20. Light diminishes in intensity as we recede from the source of light. If the luminous source be a *point*, the intensity diminishes *as the square of the distance increases.* Calling the quantity of light falling upon a given surface at the distance of a foot or a yard—1, the quantity falling on it at a distance of 2 feet or 2 yards is $\frac{1}{4}$, at a distance of 3 feet or 3 yards it is $\frac{1}{9}$, at a distance of 10 feet or 10 yards it would be $\frac{1}{100}$, and so on. This is the meaning of the law of inverse squares as applied to light.

21. *Experimental Illustrations.*—Place your source of light, which may be a candle-flame—though the law is in strictness true only for *points*—at a distance say of 9 feet from a white screen. Hold a square of pasteboard, or some other suitable material, at a

distance of $2\frac{1}{4}$ feet from the flame, or $\frac{1}{4}$th of the distance of the screen. The square casts a shadow upon the screen.

22. Assure yourself that the area of this shadow is sixteen times that of the square which casts it; a student of Euclid will see in a moment that this must be the case, and those who are not geometers can readily satisfy themselves by actual measurement. Dividing, for example, each side of a square sheet of paper into four equal parts, and folding the sheet at the opposite points of division, a small square is obtained $\frac{1}{16}$th of the area of the large one. Let this small square, or one equal to it, be your shadow-casting body. Held at $2\frac{1}{4}$ feet from the flame, its shadow upon the screen 9 feet distant will be exactly covered by the entire sheet of paper. When therefore the small square is removed, the light that fell upon it is diffused over sixteen times the area on the screen; it is therefore diluted to $\frac{1}{16}$th of its former intensity. That is to say, by augmenting the distance four-fold we diminish the light sixteen-fold.

23. Make the same experiment by placing a square at a distance of 3 feet from the source of light and 6 from the screen. The shadow now cast by the square will have nine times the area of the square itself; hence the light falling on the square is diffused over nine times the surface upon the screen. It is therefore reduced to $\frac{1}{9}$th of its intensity. That is to say, by trebling the distance from the source of light we diminish the light nine-fold.

24. Make the same experiment at a distance of $4\frac{1}{2}$ feet from the source. The shadow here will be four times the area of the shadow-casting square, and the light diffused over the greater square will be reduced to $\frac{1}{4}$th of its former intensity. Thus, by doubling the distance from the source of light we reduce the intensity of the light four-fold.

25. Instead of beginning with a distance of $2\frac{1}{4}$ feet from the source, we might have begun with a distance of 1 foot. The area of the shadow in this case would be eighty-one times that of the square which casts it; proving that at 9 feet distance the intensity of the light is $\frac{1}{81}$ of what it is at 1 foot distance.

26. Thus when the distances are

$$1,\quad 2,\quad 3,\quad 4,\quad 5,\quad 6,\quad 7,\quad 8,\quad 9, \quad \&c.,$$

the relative intensities are

$$1,\quad \tfrac{1}{4},\quad \tfrac{1}{9},\quad \tfrac{1}{16},\quad \tfrac{1}{25},\quad \tfrac{1}{36},\quad \tfrac{1}{49},\quad \tfrac{1}{64},\quad \tfrac{1}{81},\quad \&c.$$

This is the numerical expression of the law of inverse squares.

Photometry, or the Measurement of Light.

27. The law just established enables us to compare one light with another, and to express by numbers their relative illuminating powers.

28. The more intense a light, the darker is the shadow which it

casts; in other words, the greater is the contrast between the illuminated and unilluminated surface.

29. Place an upright rod in front of a white screen and a candle-flame at some distance behind the rod, the rod casts a shadow upon the screen.

30. Place a second flame by the side of the first, a second shadow is cast, and it is easy to arrange matters so that the shadows shall be close to each other, thus offering themselves for easy comparison to the eye. If when the lights are at the same distance from the screen the two shadows are equally dark, then the two lights have the same illuminating power.

31. But if one of the shadows be darker than the other, it is because its corresponding light is brighter than the other. Remove the brighter light farther from the screen, the shadows gradually approximate in depth, and at length the eye can perceive no difference between them. The shadow corresponding to each light is now illuminated by the other light, and if the shadows are equal it is because the quantities of light cast by both upon the screen are equal.

32. Measure the distances of the two lights from the screen, and square these distances. The two squares will express the relative illuminating powers of the two lights. Supposing one distance to be 3 feet and the other 5, the relative illuminating powers are as 9 to 25.

Brightness.

33. But if light diminishes so rapidly with the distance—if, for example, the light of a candle at the distance of a yard is 100 times more intense than at the distance of 10 yards—how is it that on looking at lights in churches or theatres, or in large rooms, or at our street lamps, a light 10 yards off appears almost, if not quite, as bright as one close at hand ?

34. To answer this question I must anticipate matters so far as to say that at the back of the eye is a screen, woven of nerve-filaments, named the retina; and that when we see a light distinctly, its image is formed upon this screen. This point will be fully developed when we come to treat of the eye. Now the sense of external brightness depends upon the brightness of this internal retinal image, and not upon its size. As we retreat from a light, its image upon the retina becomes smaller, and it is easy to prove that the diminution follows the law of inverse squares. That at a double distance the area of the retinal image is reduced to one-fourth, at a treble distance to one-ninth, and so on. The concentration of light accompanying this decrease of magnitude exactly atones for the diminution due to distance; hence, if the air be clear, the light, within wide variations of distance, appears equally bright to the observer.

35. If an eye could be placed behind the retina, the augmentation

or diminution of the image, with the decrease or increase of distance, might be actually observed. An exceedingly simple apparatus enables us to illustrate this point. Take a pasteboard or tin tube, three or four inches wide and three or four inches long, and cover one end of it with a sheet of tinfoil, and the other end with tracing-paper, or ordinary letter paper wetted with oil or turpentine. Prick the tinfoil with a needle, and turn the aperture towards a candle-flame. An inverted image of the flame will be seen on the translucent paper screen by the eye behind it. As you approach the flame the image becomes larger, as you recede from the flame the image becomes smaller; but the *brightness* remains throughout the same. It is so with the image upon the retina.

36. If a sunbeam be permitted to enter a room through a small aperture, the spot of light formed on a distant screen will be *round*, whatever be the shape of the aperture; this curious effect is due to the angular magnitude of the sun. Were the sun a *point*, the light spot would be accurately of the same shape as the aperture. Supposing then the aperture to be square, every point of light round the sun's periphery sends a small square to the screen. These small squares are ranged round a circle corresponding to the periphery of the sun; through their blending and overlapping they produce a rounded outline. The spots of light which fall through the apertures of a tree's foliage on the ground are rounded for the same reason.

Light requires Time to pass through Space.

37. This was proved in 1675 and 1676 by an eminent Dane, named Olaf Rœmer, who was then engaged with Cassini in Paris in observing the eclipses of Jupiter's moons. The planet, whose distance from the sun is 475,693,000 miles, has four satellites. We are now only concerned with the one nearest to the planet. Rœmer watched this moon, saw it move round in front of the planet, pass to the other side of it, and then plunge into Jupiter's shadow, behaving like a lamp suddenly extinguished: at the other edge of the shadow he saw it reappear like a lamp suddenly lighted. The moon thus acted the part of a signal light to the astronomer, which enabled him to tell exactly its time of revolution. The period between two successive lightings up of the lunar lamp gave this time. It was found to be 42 hours, 28 minutes, and 35 seconds.

38. This observation was so accurate, that having determined the moment when the moon emerged from the shadow, the moment of its hundredth appearance could also be determined. In fact it would be 100 times 42 hours, 28 minutes, 35 seconds, from the first observation.

39. Rœmer's first observation was made when the earth was in the part of its orbit nearest Jupiter. About six months afterwards, when the little moon ought to make its appearance for the hundredth time, *it was found* unpunctual, being fully 15 minutes behind its calculated

time. Its appearance, moreover, had been growing gradually later, as the earth retreated towards the part of its orbit most distant from Jupiter.

40. Rœmer reasoned thus:—' Had I been able to remain at the other side of the earth's orbit, the moon might have appeared always at the proper instant; an observer placed there would probably have seen the moon 15 minutes ago, the retardation in my case being due to the fact that the light requires 15 minutes to travel from the place where my first observation was made to my present position.'

41. This flash of genius was immediately succeeded by another. ' If this surmise be correct,' Rœmer reasoned, ' then as I approach Jupiter along the other side of the earth's orbit, the retardation ought to become gradually less, and when I reach the place of my first observation there ought to be no retardation at all.' He found this to be the case, and thus proved not only that light required time to pass through space, but also determined its rate of propagation.

42. The velocity of light as determined by Rœmer is 192,500 miles in a second.

The Aberration of Light.

The astounding velocity assigned to light by the observations of Rœmer received the most striking confirmation from the English astronomer Bradley in the year 1723. In Kew Gardens to the present hour there is a sundial to mark the spot where Bradley discovered the aberration of light.

43. If we move quickly through a rain-shower which falls vertically downwards, the drops will no longer seem to fall vertically, but will appear to meet us. A similar deflection of the stellar rays by the motion of the earth in its orbit is called the *aberration of light*.

44. Knowing the speed at which we move through a vertical rain-shower, and knowing the angle at which the rain-drops appear to descend, we can readily calculate the velocity of the falling drops of rain. So likewise, knowing the velocity of the earth in its orbit, and the deflection of the rays of light produced by the earth's motion, we can immediately calculate the velocity of light.

45. The velocity of light, as determined by Bradley, is 191,515 miles per second—a most striking agreement with the result of Rœmer.

46. This velocity has also been determined by experiments over terrestrial distances. M. Fizeau found it thus to be 194,677 miles a second, while the later experiments of M. Foucault made it 185,177 miles a second.

47. ' A cannon-ball,' says Sir John Herschel, ' would require seventeen years to reach the sun, yet light travels over the same space in eight minutes. The swiftest bird, at its utmost speed, would require nearly three weeks to make the tour of the earth. Light performs the same distance in much less time than is necessary for a

single stroke of its wing ; yet its rapidity is but commensurate with the distance it has to travel. It is demonstrable that light cannot reach our system from the nearest of the fixed stars in less than five years, and telescopes disclose to us objects probably many times more remote.'

The Reflexion of Light (Catoptrics)—Plane Mirrors.

48. When light passes from one optical medium to another, a portion of it is always turned back or reflected.

49. Light is *regularly* reflected by a polished surface ; but if the surface be not polished the light is *irregularly* reflected or scattered.

50. Thus a piece of ordinary drawing-paper will scatter a beam of light that falls upon it so as to illuminate a room. A plane mirror receiving the sunbeam will reflect it in a definite direction, and illuminate intensely a small portion of the room.

51. If the polish of the mirror were perfect it would be invisible, we should simply see in it the images of other objects ; if the room were without dust particles, the beam passing through the air would also be invisible. It is the light scattered by the mirror and by the particles suspended in the air which renders them visible.

52. A ray of light striking as a perpendicular against a reflecting surface is reflected back along the perpendicular ; it simply retraces its own course. If it strike the surface obliquely, it is reflected obliquely.

53. Draw a perpendicular to the surface at the point where the ray strikes it ; the angle enclosed between the *direct* ray and this perpendicular is called the angle of incidence. The angle enclosed by the *reflected* ray and the perpendicular is called the angle of reflexion.

54. It is a fundamental law of optics that *the angle of incidence is equal to the angle of reflexion.*

Verification of the Law of Reflexion.

55. Fill a basin with water to the brim, the water being blackened by a little ink. Let a small plummet—a small lead bullet, for example—suspended by a thread, hang into the water. The water is to be our horizontal mirror, and the plumb-line our perpendicular. Let the plummet hang from the centre of a horizontal scale, with inches marked upon it right and left from the point of suspension, which is to be the zero of the scale. A lighted candle is to be placed on one side of the plumb-line, the observer's eye being at the other.

56. The question to be solved is this :—How is the ray which strikes the liquid surface at the foot of the plumb-line reflected ? Moving the candle along the scale, so that the tip of its flame shall stand opposite different numbers, it is found that, to see the reflected tip of the flame *in the direction of the foot of the plumb-line*, the line *of vision* must cut the scale as far on the one side of that line as the

candle is on the other. In other words, the ray reflected from the foot of the perpendicular cuts the scale accurately at the candle's distance on the other side of the perpendicular. From this it immediately follows that the angle of incidence is equal to the angle of reflexion.

57. With an artificial horizon of this kind, and employing a theodolite to take the necessary angles, the law has been established with the most rigid accuracy. The angle of elevation to a star being taken by the instrument, the telescope is then pointed downwards to the image of the star reflected from the artificial horizon. It is always found that the direct and reflected rays enclose equal angles with the horizontal axis of the telescope, the reflected ray being as far below the horizontal axis as the direct ray is above it. On account of the star's distance the ray which strikes the reflecting surface is parallel with the ray which reaches the telescope directly, and from this follows, by a brief but rigid demonstration, the law above enunciated.

58. The path described by the direct and reflected rays is the shortest possible.

59. When the reflecting surface is roughened, rays from different points, more or less distant from each other, reach the eye. Thus, a breeze crisping the surface of the Thames or Serpentine sends to the eye, instead of single images of the lamps upon their margin, pillars of light. Blowing upon our basin of water, we also convert the reflected light of our candle into a luminous column.

60. Light is reflected with different energy by different substances. At a perpendicular incidence, only 18 rays out of every 1000 are reflected by water, 25 rays per 1000 by glass, while 666 per 1000 are reflected by mercury.

61. When the rays strike obliquely, a greater amount of light than that stated in 60 is reflected by water and glass. Thus, at an incidence of 40°, water reflects 22 rays; at 60°, 65 rays; at 80°, 333 rays; and at $89\frac{1}{2}$° (almost grazing the surface), it reflects 721 rays out of every 1000. This is as much as mercury reflects at the same incidence.

62. The augmentation of the light reflected as the obliquity of incidence is increased may be illustrated by our basin of water. Hold the candle so that its rays enclose a large angle with the liquid surface, and notice the brightness of its image. Lower both the candle and the eye until the direct and reflected rays as nearly as possible graze the liquid surface; the image of the flame is now much brighter than before.

Reflexion from Looking-glasses.—Various instructive experiments with a looking-glass may here be performed and understood.

63. Note first when a candle is placed between the glass and the eye, so that a line from the eye through the candle is perpendicular to the glass, that *one* well-defined image of the candle only is seen.

64. Let the eye now be moved so as to receive an oblique reflexion; the image is no longer single, a series of images at first partially

overlapping each other being seen. By rendering the incidence sufficiently oblique these images, if the glass be sufficiently thick, may be completely separated from each other.

65. The first image of the series arises from the reflexion of the light from the anterior surface of the glass.

66. The second image, which is usually much the brightest, arises from reflexion at the silvered surface of the glass. At large incidences, as we have just learned, metallic reflexion far transcends that from glass.

67. The other images of the series are produced by the reverberation of the light from surface to surface of the glass. At every return from the silvered surface a portion of the light quits the glass and reaches the eye, forming an image; a portion is also sent back to the silvered surface, where it is again reflected. Part of this reflected beam also reaches the eye and yields another image. This process continues: the quantity of light reaching the eye growing gradually less, and, as a consequence, the successive images growing dimmer, until finally they become too dim to be visible.

68. A very instructive experiment illustrative of the augmentation of the reflexion from glass, through augmented obliquity, may here be made. Causing the candle and the eye to approach the looking-glass, the first image becomes gradually brighter; and you end by rendering the image reflected from the glass brighter, more luminous, than that reflected from the metal. Irregularities in the reflexion from looking-glasses often show themselves; but with a good glass—and there are few glasses so defective as not to possess, at all events, some good portions—the succession of images is that here indicated.

69. *Position and Character of Images in Plane Mirrors.*—The image in a plane mirror appears as far behind the mirror as the object is in front of it. This follows immediately from the law which announces the equality of the angles of incidence and reflexion. Draw a line representing the section of a plane mirror; place a point in front of it. Rays issue from that point, are reflected from the mirror, and strike the pupil of the eye. The pupil is the base of a cone of such rays. Produce the rays backward; they will intersect behind the mirror, and the point will be seen *as if* it existed at the place of intersection. The place of intersection is easily proved to be as far behind the mirror as the point is in front of it.

70. Exercises in determining the positions of images in a plane mirror, the positions of the objects being given, are here desirable. The image is always found by simply letting fall a perpendicular from each point of the object, and producing it behind the mirror, so as to make the part behind equal to the part in front. We thus learn that the image is of the same size and shape as the object, agreeing with it in all respects save one—the image is a *lateral inversion* of the object.

71. *This* inversion enables us, by means of a mirror, to read

writing written backward, as if it were written in the usual way. Compositors arrange their type in this backward fashion, the type being reversed by the process of printing. A looking-glass enables us to read the type as the printed page.

72. Lateral inversion comes into play when we look at our own faces in a glass. The right cheek of the object, for example, is the left cheek of the image; the right hand of the object the left hand of the image, &c. The hair parted on the left in the object is seen parted to the right of the image, &c.

73. A plane mirror half the height of an object gives an image which embraces the whole height. This is readily deduced from what has gone before.

74. If a plane mirror be caused to move parallel with itself, the motion of an image in the mirror moves with twice its rapidity.

75. The same is true of a *rotating* mirror: when a plane mirror is caused to rotate, the angle described by the image is twice that described by the mirror.

76. In a mirror inclined at an angle of 45 degrees to the horizon, the image of an erect object appears horizontal, while the image of a horizontal object appears erect.

77. An object placed between two mirrors enclosing an angle yields a number of images depending upon the angle enclosed by the mirrors. The smaller the angle, the greater is the number of images. To find the number of images, divide 360° by the number of degrees in the angle enclosed by the two mirrors, the quotient, if a whole number, will be the number of images, plus one, or it will include the images and the object. The construction of the kaleidoscope depends on this.

78. When the angle becomes 0,—in other words, when the mirrors are parallel,—the number of images is infinite. Practically, however, we see between parallel mirrors a long succession of images, which become gradually feebler, and finally cease to be sensible to the eye.

Reflexion from Curved Surfaces: Concave Mirrors.

79. It has been already stated and illustrated that light moves in straight lines, which receive the name of rays. Such rays may be either divergent, parallel, or convergent.

80. Rays issuing from terrestrial points are necessarily divergent. Rays from the sun or stars are, in consequence of the immense distances of these objects, sensibly parallel.

81. By suitably reflecting them, we can render the rays from terrestrial sources either parallel or convergent. This is done by means of *concave* mirrors.

82. In its reflexion from such mirrors, light obeys the law already enunciated for plane mirrors. The angle of incidence is equal to the angle of reflexion.

83. Let M N be a very small portion of the circumference of a circle

with its centre at O. Let the line $a\,x$, passing through the centre, cut the arc M N into two equal parts at a. Then imagine the curve M N twirled round $a\,x$ as a fixed axis; the curve would describe part of a spherical surface. Suppose the surface turned towards x to be silvered over, we should then have a concave spherical reflector; and we have now to understand the action of this reflector upon light.

Fig. 1.

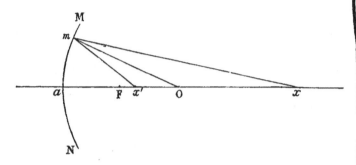

84. The line $a\,x$ is the principal axis of the mirror.

85. All rays from a point placed at the centre O strike the surface of the mirror as perpendiculars, and after reflexion return to O.

86. A luminous point placed on the axis beyond O, say at x, throws a divergent cone of rays upon the mirror. These rays are rendered convergent on reflexion, and they intersect each other at some point on the axis between the centre O and the mirror. In every case the direct and the reflected rays ($x\,m$ and $m\,x'$ for example) enclose equal angles with the radius (O m) drawn to the point of incidence.

87. Supposing x to be exceedingly distant, say as far away as the sun from the small mirror,—or, more correctly, supposing it to be *infinitely* distant,—then the rays falling upon the mirror will be *parallel.* After reflexion such rays intersect each other, *at a point midway between the mirror and its centre.*

88. This point, which is marked F in the figure, is the *principal focus* of the mirror; that is to say, the principal focus is the focus of *parallel rays.*

89. The distance between the surface of the mirror and its principal focus is called the *focal distance.*

90. In optics, the position of an object and of its image are always exchangeable. If a luminous point be placed in the principal focus, the rays from it will, after reflexion, be parallel. If the point be placed anywhere between the principal focus and the centre O, the rays after reflexion will cut the axis at some point beyond the centre.

91. If the point be placed between the principal focus F and the

mirror, the rays after reflexion will be *divergent*—they will not intersect at all—there will be no *real* focus.

92. But if these divergent rays be produced backwards, they will intersect *behind* the mirror, and form there what is called a *virtual*, or imaginary focus.

Before proceeding further, it is necessary that these simple details should be thoroughly mastered. Given the position of a point in the axis of a concave mirror, no difficulty must be experienced in finding the position of the image of that point, nor in determining whether the focus is *virtual* or *real*.

93. It will thus become evident that while a point moves from an infinite distance to the centre of a spherical mirror, the image of that point moves only over the distance between the principal focus and the centre. Conversely, it will be seen that during the passage of a luminous point from the centre to the principal focus, the image of the point moves from the centre to an infinite distance.

94. The point and its image occupy what are called *conjugate foci*. If the last note be understood, it will be seen that the conjugate foci move in opposite directions, and that they coincide at the centre of the mirror.

95. If instead of a point an object of sensible dimensions be placed beyond the centre of the mirror, an *inverted* image of the object *diminished* in size will be formed between the centre and the principal focus.

96. If the object be placed between the centre and the principal focus, an inverted and *magnified* image of the object will be formed beyond the centre. The positions of the image and its object are, it will be remembered, convertible.

97. In the two cases mentioned in 95 and 96 the image is formed in the air in *front* of the mirror. It is a *real* image. But if the object be placed between the principal focus and the mirror, an *erect* and magnified image of the object is seen behind the mirror. The image is here *virtual*. The rays enter the eye *as if* they came from an object behind the mirror.

98. It is plain that the images seen in a common looking-glass are all virtual images.

99. It is now to be noted that what has been here stated regarding the gathering of rays to a *single focus* by a spherical mirror is only true when the mirror forms a small fraction of the spherical surface. Even then it is only practically, not strictly and theoretically, true.

Caustics by Reflexion (Catacaustics).

100. When a large fraction of the spherical surface is employed as a mirror, the rays are not all collected to a point; their intersections,

on the contrary, form a luminous *surface*, which in optics is called a *caustic* (German, Brennfläche).

101. The interior surface of a common drinking-glass is a curved reflector. Let the glass be nearly filled with milk, and a lighted candle placed beside it; a caustic curve will be drawn upon the surface of the milk. A carefully bent hoop, silvered within, also shows the caustic very beautifully. The focus of a spherical mirror is the *cusp* of its caustic.

102. *Aberration.*—The deviation of any ray from this cusp is called the *aberration* of the ray. The inability of a spherical mirror to collect all the rays falling upon it to a single point is called the *spherical aberration* of the mirror.

103. Real images, as already stated, are formed in the air in front of a concave mirror, and they may be seen in the air by an eye placed among the divergent rays beyond the image. If an opaque screen, say of thick paper, intersect the image, it is projected on the screen and is seen *in all positions* by an eye placed in front of the screen. If the screen be semi-transparent, say of ground glass or tracing-paper, the image is seen by an eye placed either in front of the screen or behind it. The images in phantasmagoria are thus formed.

Concave spherical surfaces are usually employed as burning-mirrors. By condensing the sunbeams with a mirror 3 feet in diameter and of 2 feet focal distance, very powerful effects may be obtained. At the focus, water is rapidly boiled, and combustible bodies are immediately set on fire. Thick paper bursts into flame with explosive violence, and a plank is pierced as with a hot iron.

Convex Mirrors.

104. In the case of a *convex* spherical mirror the positions of its foci and of its images are found as in the case of a concave mirror. But all the foci and all the images of a convex mirror are virtual.

105. Thus to find the principal focus you draw parallel rays which, on reflection, enclose angles with the radii equal to those enclosed by the direct rays. The reflected rays are here *divergent*; but on being produced backwards, they intersect at the principal focus *behind the mirror*.

106. The drawing of *two* lines suffices to fix the position of the image of any point of an object either in concave or convex spherical mirrors. A ray drawn from the point through the centre of the mirror will be reflected through the centre; a ray drawn parallel to the axis of the mirror will, after reflection, pass, or its production will pass, through the principal focus. The intersection of these two reflected rays determines the position of the image of the point. Applying this construction to objects of sensible magnitude, it follows that the image of an object in a convex mirror is always *erect* and *diminished*.

107. If the mirror be *parabolic* instead of spherical, all parallel

rays falling upon the mirror are collected to a point at its focus; conversely, a luminous point placed at the focus sends forth parallel rays: there is no aberration. If the mirror be *elliptical*, all rays emitted from one of the foci of the ellipsoid are collected together at the other. Parabolic reflectors are employed in lighthouses, where it is an object to send a powerful beam, consisting of rays as nearly as possible parallel, far out to sea. In this case the centre of the flame is placed in the focus of the mirror; but, inasmuch as the flame is of sensible magnitude, and not a mere point, the rays of the reflected beam are not accurately parallel.

The Refraction of Light (*Dioptrics*).

108. We have hitherto confined our attention to the portion of a beam of light which rebounds from the reflecting surface. But in general, a portion of the beam also *enters* the reflecting substance, being rapidly quenched when the substance is opaque (see note 11), and freely *transmitted* when the substance is transparent.

109. Thus in the case of water, mentioned in note 60, when the incidence is perpendicular all the rays are transmitted, save the 18 referred to as being reflected. That is to say, 982 out of every 1000 rays enter the water and pass through it.

110. So likewise in the case of mercury, mentioned in the same note; 334 out of every 1000 rays falling on the mercury at a perpendicular incidence, enter the metal and are quenched at a minute depth beneath its surface.

We have now to consider that portion of the luminous beam which enters the reflecting substance; taking, as an illustrative case, the passage from air into water.

111. If the beam fall upon the water as a perpendicular, it pursues a straight course through the water: if the incidence be oblique, the direction of the beam is changed at the point where it enters the water.

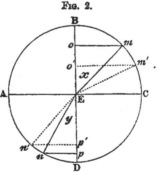

Fig. 2.

112. This bending of the beam is called *refraction*. Its amount is different in different substances.

113. The refraction of light obeys a perfectly rigid law which must be clearly understood. Let A B C D, fig. 2, be the section of a cylindrical vessel which is half filled with water, its surface being A C. E is the centre of the circular section of the cylinder, and B D is a perpendicular to the surface at E. Let the cylindrical envelope of the vessel be opaque, say of brass or tin, and let an aperture be imagined in it at B, through

which a narrow light-beam passes to the point E. The beam will pursue a straight course to D without turning to the right or to the left.

114. Let the aperture be imagined at *m*, the beam striking the surface of the water at E *obliquely*. Its course on entering the liquid will be changed; it will pursue the track E *n*.

115. Draw the line *m o* perpendicular to B D, and also the line *n p* perpendicular to the same B D. It is always found that *m o* divided by *n p* is *a constant quantity*, no matter what may be the angle at which the ray enters the water.

116. The angle marked *x* above the surface is called the angle of incidence; the angle at *y* below the surface is called the angle of refraction; and if we regard the radius of the circle A B C D as unity or 1, the line *m o* will be the *sine* of the angle of incidence; while the line *n p* will be the *sine* of the angle of refraction.

117. Hence the all-important optical law—*The sine of the angle of incidence divided by the sine of the angle of refraction is a constant quantity.* However these angles may vary in size, this bond of relationship is never severed. If one of them be lessened or augmented, the other must diminish or increase so as to obey this law. Thus if the incidence be along the dotted line *m'* E, the refraction will be along the line E *n'*, but the ratio of *m' o* to *n' p'* will be precisely the same as that of *m o* to *n p*.

118. The constant quantity here referred to is called *the index of refraction.*

119. One word more is necessary to the full comprehension of the term sine, and of the experimental demonstration of the law of refraction. When one number is divided by another the quotient is called the *ratio* of the one number to the other. Thus 1 divided by 2 is $\frac{1}{2}$, and this is the ratio of 1 to 2. Thus also 2 divided by 1 is 2, and this is the ratio of 2 to 1. In like manner 12 divided by 3 is 4, and this is the ratio of 12 to 3. Conversely, 3 divided by 12 is $\frac{1}{4}$, and this is the ratio of 3 to 12.

120. In a right-angled triangle the ratio of any side to the hypothenuse is found by dividing that side by the hypothenuse. *This ratio is the sine of the angle opposite to the side*, however large or small the triangle may be. Thus in fig. 2 the sine of the angle *x* in the right-angled triangle E *o m* is really the ratio of the line *o m* to the hypothenuse E *m*; it would be expressed in a fractional form thus, $\frac{o\,m}{E\,m}$. In like manner the sine of *y* is the ratio of the line *n p* to the hypothenuse E *n*, and would be expressed in a fractional form thus, $\frac{n\,p}{E\,n}$. These fractions are the sines of the respective angles, whatever be the length of the line E *m* or E *n*. In the particular *case above* referred to, where these lines are considered as units, the

fractions $\frac{m\,o}{1}$ and $\frac{n\,p}{1}$, or in other words $m\,o$ and $n\,p$, become, as stated, the sines of the respective angles. We are now prepared to understand a simple but rigid demonstration of the law of refraction.

FIG. 3.

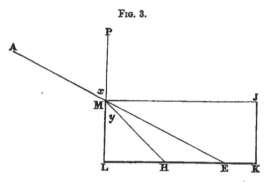

121. M L J K is a cell with parallel glass sides and one opaque end M L. The light of a candle placed at A falls into the vessel, the end M L casting a shadow which reaches to the point E. Fill the vessel with water,—the shadow retreats to H through the refraction of the light at the point where it enters the water.

122. The angle enclosed between M E and M L is equal to the angle of incidence x, and in accordance with the definition given in 120, $\frac{L\,E}{M\,E}$ is its sine; while $\frac{L\,H}{M\,H}$ is the sine of the angle of refraction y. All these lines can be either measured or calculated. If they be thus determined, and if the division be actually made, it will always be found that the two quotients $\frac{L\,E}{M\,E}$ and $\frac{L\,H}{M\,H}$ stand in a constant ratio to each other, whatever the angle may be at which the light from A strikes the surface of the liquid. This ratio in the case of water is $\frac{4}{3}$, or, expressed in decimals, 1·333.*

123. When the light passes from air into water, the refracted ray is bent *towards* the perpendicular. This is generally, but not always, the case when the light passes from a rarer to a denser medium.

124. The principle of reversibility which runs through the whole of optics finds illustration here. When the ray passes from water to air it is bent *from* the perpendicular: it accurately reverses its course.

125. If instead of water we employed vinegar the ratio would be 1·344; with brandy it would be 1·360; with rectified spirit of wine 1·372; with oil of almonds or with olive oil 1·470; with spirit of

* More accurately, 1·336.

c

turpentine 1·605; with oil of aniseed 1·538; with oil of bitter almonds 1·471; with bisulphide of carbon 1·678; with phosphorus 2·24.

126. These numbers express the indices of refraction of the various substances mentioned; all of them refract the light more powerfully than water, and it is worthy of remark that all of them, except vinegar, are *combustible* substances.

127. It was the observation on the part of Newton, that, having regard to their density, 'unctuous substances' as a general rule refracted light powerfully, coupled with the fact that the index of refraction of the diamond reached, according to his measurements, so high a figure as 2·439, that caused him to foresee the possible combustible nature of the diamond. The bold prophecy of Newton* has been fulfilled, the combustion of a diamond being one of the commonest experiments of modern chemistry.

128. It is here worth noting that the refraction by spirit of turpentine is greater than that by water, though the density of the spirit is to that of the water as 874 is to 1000. A ray passing obliquely from the spirit of turpentine into water is bent *from* the perpendicular, though it passes from a rarer to a denser medium; while a ray passing from water into the spirit of turpentine is bent *towards* the perpendicular, though it passes from a denser to a rarer medium. Hence the necessity for the words 'not always' employed in 123.

129. If a ray of light pass through a refracting plate with parallel surfaces, or through any number of plates with parallel surfaces, on regaining the medium from which it started, its original direction is restored. This follows from the principle of reversibility already referred to.

130. In passing through a refracting body, or through any number of refracting bodies, the light accomplishes its transit in the *minimum of time*. That is to say, given the velocity of light in the various media, the path chosen by the ray, or, in other words, the path which its refraction imposes upon the ray, enables it to perform its journey in the most rapid manner possible.

131. Refraction always causes water to appear shallower, or a transparent plate of any kind thinner, than it really is. The lifting up of the lower surface of a glass cube, through this cause, is very remarkable.

132. To understand why the water appears shallower, fix your attention on a point at its bottom, and suppose the line of vision from that point to the eye to be perpendicular to the surface of the water. Of all rays issuing from the point, the perpendicular one alone reaches the eye without refraction. Those close to the perpendicular, on emerging from the water, have their divergence augmented

* 'Car ce grand homme, qui mettait la plus grande sévérité dans ses expériences, et la plus grande réserve dans ses conjectures, n'hésitait jamais à suivre *les conséquences* d'une vérité aussi loin qu'elle pouvait le conduire.'—BIOT.

by refraction. Producing these divergent rays backwards, they intersect at a point above the real bottom, and at this point the bottom will be seen.

133. The apparent shallowness is augmented by looking obliquely into the water.

134. In consequence of this apparent rise of the bottom, a straight stick thrust into water is bent at the surface *from* the perpendicular.

Note the difference between the deportment of the stick and of a luminous beam. The beam on entering the water is bent *towards* the perpendicular.

135. This apparent lifting of the bottom when water is poured into a basin brings into sight an object at the bottom of the basin which is unseen when the basin is empty.

Opacity of Transparent Mixtures.

136. Reflexion always accompanies refraction; and if one of these disappear, the other will disappear also. A solid body immersed in a liquid having the same refractive index as the solid, vanishes; it is no more seen than a portion of the liquid itself of the same size would be seen.

137. But in the passage from one medium to another of a different refractive index, light is always reflected; and this reflexion may be so often repeated as to render the mixture of two transparent substances practically impervious to light. It is the frequency of the reflexions at the limiting surfaces of air and water that renders *foam* opaque. The blackest clouds owe their gloom to this repeated reflexion, which diminishes their *transmitted* light. Hence also their whiteness by *reflected* light. To a similar cause is due the whiteness and imperviousness of common salt, and of transparent bodies generally when crushed to powder. The individual particles transmit light freely; but the reflexions at their surfaces are so numerous that the light is wasted in echoes before it can reach to any depth in the powder.

138. The whiteness and opacity of writing-paper are due mainly to the same cause. It is a web of transparent fibres, not in optical contact, which intercept the light by repeatedly reflecting it.

139. But if the interstices of the fibres be filled by a body of the same refractive index as the fibres themselves, the reflexion at their limiting surfaces is destroyed, and the paper is rendered transparent. This is the philosophy of the tracing-paper used by engineers. It is saturated with some kind of oil, the lines of maps and drawings being easily copied *through it* afterwards. Water augments the transparency of paper, as it darkens a white towel; but its refractive index is too low to confer on either any high degree of transparency. It however renders certain minerals, which are opaque when dry, translucent.

140. The higher the refractive index the more copious is the reflexion. The refractive index of water, for example, is 1·336; that

of glass is 1·5. Hence the different quantities of light reflected by water and glass at a perpendicular incidence, as mentioned in note 60. It is its enormous refractive strength that confers such brilliancy upon the diamond.

Total Reflexion.

Read notes 123 and 124; then continue here.

141. When the angle of incidence from air into water is nearly 90°, that is to say, when the ray before entering the water just grazes its surface, the angle of refraction is $48\frac{1}{2}$°. Conversely, when a ray passing from water into air strikes the surface at an angle of $48\frac{1}{2}$° it will, on its emergence, just graze the surface of the water.

142. If the angle which the ray in water encloses with the perpendicular to the surface be greater than $48\frac{1}{2}$°, the ray will not quit the water at all: it will be *totally reflected* at the surface.

143. The angle which marks the limit where total reflexion begins is called the *limiting angle* of the medium. For water this angle is 48° 27′, for flint glass it is 38° 41′, while for diamond it is 23° 42′.

144. Realise clearly that a bundle of light rays filling an angular space of 90° before they enter the water, are squeezed into an angular space of 48° 27′ within the water, and that in the case of diamond the condensation is from 90° to 23° 42′.

145. To an eye in still water its margin must appear *lifted up.* A fish, for example, sees objects, as it were, through a circular aperture of about 97° (twice 47° 27′) in diameter overhead. All objects down to the horizon will be visible in this space, and those near the horizon will be much distorted and contracted in dimensions, especially in height. Beyond the limits of this circle will be seen the bottom of the water totally reflected, and therefore depicted as vividly as if seen by direct vision.*

146. A similar effect, exerted by the atmosphere (when no clouds cross the orbs), gives the sun and moon at rising and setting a slightly flattened appearance.

147. *Experimental Illustrations.*—Place a shilling in a drinking-glass; cover it with water to about the depth of an inch, and tilt the glass so as to obtain the necessary obliquity of incidence at the surface. Looking upwards towards the surface, the image of the shilling will be seen shining there, and as the reflexion is total, the image will be as bright as the shilling itself. A spoon suitably dipped into the glass also yields an image due to total reflexion.

148. Thrust the closed end of an empty test-tube into a glass of water. Incline the tube, until the horizontal light falling upon it shall be totally reflected upwards. When looked down upon, the tube appears shining like burnished silver. Pour a little water into the

* Sir John Herschel.

tube: as the liquid rises, it abolishes total reflexion, and with it the lustre, leaving a gradually diminishing lustrous zone, which disappears wholly when the level of the water within rises to, or above, that of the water without. A tube of any kind stopped watertight will answer for this experiment, which is both beautiful and instructive.

149. If a ray of light fall as a perpendicular on the side of a right-angled isosceles glass prism, it will enter the glass and strike the hypothenuse at an angle of 45°. This exceeds the limiting angle of glass; the ray will therefore be totally reflected; and, in accordance with the law mentioned in note 54, the direct and reflected rays will be at right angles to each other. When such a change of direction is required in optical instruments, a right-angled isosceles prism is usually employed.

150. When the ray enters the prism parallel to the hypothenuse, it will be refracted, and will strike the hypothenuse at an angle greater than the limiting angle. It will therefore be totally reflected. If the object, instead of being a point, be of sensible magnitude, the rays from its extremities will *cross each other* within the prism, and hence the object will appear *inverted* when looked at through the prism. Dove has applied the 'reversion prism' to render erect the inverted images of the astronomical telescope.

151. The mirage of the desert and various other phantasmal appearances in the atmosphere are, in part, due to total reflexion. When the sun heats an expanse of sand, the layer of air in contact with the sand becomes lighter than the superincumbent air. The rays from a distant object, a tree for example, striking very obliquely upon the upper surface of this layer, may be totally reflected, thus showing images similar to those produced by a surface of water. The thirsty soldiers of the French army were tantalised by such appearances in Egypt.

152. Gases, like liquids and solids, can refract and reflect light; but, in consequence or the lowness of their refractive indices, both reflection and refraction are feeble. Still atmospheric refraction has to be taken into account by the astronomer, and by those engaged in trigonometrical surveys. The refraction of the atmosphere causes the sun to be seen before it actually rises, and after it actually sets.

153. The quivering of objects seen through air rising over a heated surface is due to irregular refraction, which incessantly shifts the directions of the rays of light. In the air this shifting of the rays is never entirely absent, and it is often a source of grievous annoyance to the astronomer who needs a homogeneous atmosphere.

154. The flame of a candle or of a gas-lamp, and the column of heated air above the flame; the air rising from a red-hot iron; the pouring of a heavy gas, such as carbonic acid, downwards into air; and the issue of a lighter one, such as hydrogen, upwards,—may all be made to reveal themselves by their action upon a sufficiently intense light. The transparent gases interposed between such a light and

a white screen are seen to rise like smoke upon the screen through the effects of refraction.

Lenses.

155. A lens in optics is a portion of a refracting substance such as glass, which is bounded by curved surfaces. If the surface be spherical the lens is called a spherical lens.

156. Lenses divide themselves into two classes, one of which renders parallel rays convergent, the other of which renders such rays divergent. Each class comprises three kinds of lenses, which are named as follows:—

Converging Lenses.

1. Double convex, with both surfaces convex.
2. Plano-convex, with one surface plane and the other convex.
3. Concavo-convex (Meniscus), with a concave and a convex surface, the convex surface being the most strongly curved.

Diverging Lenses.

1. Double concave, with both surfaces concave.
2. Plano-concave, with one surface plane and the other concave.
3. Convexo-concave, with a convex and a concave surface, the concave surface being the most strongly curved.

157. A straight line drawn through the centre of the lens, and perpendicular to its two convex surfaces, is the principal axis of the lens.

158. A luminous beam falling on a convex lens parallel to the axis, has its constituent rays brought to intersection at a point in the axis behind the lens. This point is the principal focus of the lens. As before, the principal focus is the focus of parallel rays.

159. The rays from a luminous point placed beyond the focus intersect at the opposite side of the lens, an image of the point being formed at the place of intersection. As the point approaches the principal focus its image retreats from it, and when the point actually reaches the principal focus, its image is at an infinite distance.

160. If the principal focus be passed, and the point come between that focus and the lens, the rays after passing through the lens will be still divergent. Producing them backwards, they will intersect on that side of the lens on which stands the luminous point. The focus here is *virtual*. A body of sensible magnitude placed between the focus and the lens would have a virtual image.

161. When an object of sensible dimensions is placed anywhere beyond the principal focus, a real image of the object will be formed in the air behind the lens. The image may be either greater or less than the object in size, but the image will always be *inverted*.

162. The positions of the image and the object are, as before, convertible.

163. In the case of concave lenses the images are always virtual.

164. A spherical lens is incompetent to bring all the rays that fall upon it to the same focus. The rays which pass through the lens near its circumference are more refracted than those which pass through the central portions, and they intersect earlier. Where perfect definition is required it is therefore usual, though at the expense of illumination, to make use of the central rays only.

165. This difference of focal distance between the central and circumferential rays is called the *spherical aberration* of the lens. A lens so curved as to bring all rays to the same focus is called *aplanatic*; a spherical lens cannot be rendered aplanatic.

166. As in the case of spherical mirrors, spherical lenses have their caustic curves and surfaces (diacaustics) formed by the intersection of the refracted rays.

Vision and the Eye.

167. The human eye is a compound lens, consisting of three principal parts: the *aqueous humour*, the *crystalline lens*, and the *vitreous humour*.

168. The aqueous humour is held in front of the eye by the *cornea*, a transparent, horny capsule, resembling a watch-glass in shape. Behind the aqueous humour, and immediately in front of the crystalline lens, is the *iris*, which surrounds the *pupil*. Then follow the lens and the vitreous humour, which last constitutes the main body of the eye. The average diameter of the human eye is 10·9 lines.*

169. When the optic nerve enters the eye from behind, it divides into a series of filaments, which are woven together to form the *retina*, a delicate network spread as a screen at the back of the eye. The retina rests upon a black pigment, which reduces to a minimum all internal reflexion.

170. By means of the iris the size of the pupil may be caused to vary within certain limits. When the light is feeble the pupil expands, when it is intense the pupil contracts; thus the quantity of light admitted into the eye is, to some extent, regulated. The pupil also diminishes when the eye is fixed upon a near object, and expands when it is fixed upon a distant one.

171. The pupil appears black; partly because of the internal black coating, but mainly for another reason. Could we illuminate the retina, and see at the same time the illuminated spot, the pupil would appear bright. But the principle of reversibility, so often spoken of, comes into play here. The light of the illuminated spot in returning outwards retraces its steps, and finally falls upon the source of illumination. Hence, to receive the returning rays, the observer's eye ought to be placed between the source and the retina. But in this position it would cut off the illumination. If the light be thrown into the eye by a mirror pierced with a small orifice, or

* A line is $\frac{1}{12}$th of an inch.

with a small portion of the silvering removed, then the eye of the observer placed behind the mirror, and looking through the orifice, may see the illuminated retina. The pupil under these circumstances glows like a live coal. This is the principle of the *Ophthalmoscope* (Augenspiegel, Helmholtz), an instrument by which the interior of the eye may be scanned, and its condition in health or disease noted.

172. In the case of albinos, or of white rabbits, the black pigment is absent, and the pupil is seen red by light which passes through the *sclerotica*, or white of the eye. When this light is cut off, the pupil of an albino appears black. In some animals the black pigment is displaced by a reflecting membrane, the *tapetum*. It is the light reflected from the tapetum which causes a cat's eye to shine in partial darkness. The light in this case is not internal, for when the darkness is *total* the cat's eyes do not shine.

173. In the camera obscura of the photographer the images of external objects formed by a convex lens are received upon a plate of ground glass, the lens being pushed in or out until the image upon the glass is sharply defined.

174. The eye is a camera obscura, with its refracting lenses, the retina playing the part of the plate of ground glass in the ordinary camera. For perfectly distinct vision it is necessary that the image upon the retina should be perfectly defined; in other words, that the rays from every point of the object looked at should be converged to a *point* upon the retina.

175. The image upon the retina is *inverted*.

Adjustment of the Eye: use of Spectacles.

176. If the letters of a book held at some distance from the eye be looked at through a gauze veil placed nearer the eye, it will be found that when the letters are seen distinctly the veil is seen indistinctly; conversely, if the veil be seen distinctly, the letters will be seen indistinctly. This demonstrates that the images of objects at different distances from the eye cannot be defined *at the same time* upon the retina.

177. Were the eye a rigid mass, like a glass lens, incapable of change of form, distinct vision would only be possible at one particular distance. We know, however, that the eye possesses a power of adjustment for different distances. This adjustment is effected, not by pushing the front of the eye backwards or forwards, but by changing the curvature of the crystalline lens.

178. The image of a candle reflected from the forward or backward surface of the lens is seen to diminish when the eye changes from distant to near vision, thus proving the curvature of the lens to be greater for near than for distant vision.

179. The principal refraction endured by rays of light in crossing *the eye occurs* at the surface of the cornea, where the passage is from

air to a much denser medium. The refraction at the cornea alone would cause the rays to intersect at a point nearly half an inch behind the retina. The convergence is augmented by the crystalline lens, which brings the point of intersection forward to the retina itself.

180. A line drawn through the centre of the cornea and the centre of the whole eye to the retina is called the axis of the eye. The length of the axis, even in youth, is sometimes too small; in other words, the retina is sometimes too near the cornea; so that the refracting part of the organ is unable to converge the rays from a luminous point so as to bring them to a point upon the retina. In old age also the refracting surfaces of the eye are slightly flattened, and thus rendered incompetent to refract the rays sufficiently. In both these cases the image would be formed *behind* the retina, instead of *upon* it, and hence the vision is indistinct.

181. The defect is remedied by holding the object at a distance from the eye, so as to lessen the divergence of its rays, or by placing in front of the eye a convex lens, which helps the eye to produce the necessary convergence. This is the use of spectacles.

182. The eye is also sometimes too long in the direction of the axis, or the curvature of the refracting surfaces may be too great. In either case the rays entering the pupil are converged so as to intersect before reaching the retina. This defect is remedied either by holding the object very close to the eye, so as to augment the divergence of its rays, thus throwing back the point of intersection; or by placing in front of the eye a concave lens, which produces the necessary divergence.

183. The eye is not adjusted at the same time for equally-distant horizontal and vertical objects. The distance of distinct vision is greater for horizontal lines than for vertical ones. Draw with ink two lines at right angles to each other, the one vertical, the other horizontal: see one of them distinctly black and sharp; the other appears indistinct, as if drawn in lighter ink. Adjust the eye for this latter line, the former will then appear indistinct. This difference in the curvature of the eye in two directions may sometimes become so great as to render the application of cylindrical lenses necessary for its correction.

The Punctum Cœcum.

184. The spot where the optic nerve enters the eye, and from which it ramifies to form the network of the retina, is insensible to the action of light. An object whose image falls upon that spot is not seen. The image of a clock-face, of a human head, of the moon, may be caused to fall upon this 'blind spot,' and when this is the case the object is not visible.

185. To illustrate this point, proceed thus:—Lay two white wafers on black paper, or two black ones on white paper, with an interval of 3 inches between them. Bring the right eye at a height of 10 or 11 inches exactly over the left-hand wafer, so that the line

joining the two eyes shall be parallel to the line joining the two wafers. Closing the left eye, and looking steadily with the right at the left-hand wafer, the right-hand one ceases to be visible. In this position the image falls upon the 'blind spot' of the right eye. If the eye be turned in the least degree to the right or left, or if the distance between it and the paper be augmented or diminished, the wafer is immediately seen. Preserving these proportions as to size and distance, objects of far greater dimensions than the wafer may have their images thrown upon the blind spot, and be obliterated.

Persistence of Impressions.

186. An impression of light once made upon the retina does not subside instantaneously. An electric spark is sensibly instantaneous; but the impression it makes upon the eye remains for some time after the spark has passed away. This interval of persistence varies with different persons, and amounts to a sensible fraction of a second.

187. If, therefore, a succession of sparks follow each other at intervals less than the time which the impression endures, the separate impressions will unite to form *a continuous* light. If a luminous point be caused to describe a circle in less than this interval, the circle will appear as a continuous closed curve. From this cause also, the spokes of a rapidly rotating wheel blend together to a shadowy surface. Wheatstone's Photometer is based on this persistence. It also explains the action of those instruments in which a series of objects in different positions being brought in rapid succession before the eye, the impression of *motion* is produced.

188. A jet of water descending from an orifice in the bottom of a vessel exhibits two distinct parts: a tranquil pellucid portion near the orifice, and a turbid or untranquil portion lower down. Both parts of the jet appear equally continuous. But when the jet in a dark room is illuminated by an electric spark, all the turbid portion reveals itself as a string of separate drops standing perfectly still. It is their quick succession that produces the impression of continuity. The most rapid cannon-ball, illuminated by a flash of lightning, would be seen for the fraction of a second perfectly motionless in the air.

189. The eye is by no means a perfect optical instrument. It suffers from spherical aberration; a scattered luminosity, more or less strong, always surrounding the defined images of luminous objects upon the retina. By this luminosity the image of the object is sensibly increased in size; but with ordinary illumination the scattered light is too feeble to be noticed. When, however, bodies are intensely illuminated, more especially when the bodies are small, so that a slight extension of their images upon the retina becomes noticeable, such bodies appear larger than they really are. Thus, a platinum-wire *raised to whiteness* by a voltaic current has its apparent diameter *enormously* increased. Thus also the crescent moon seems to belong

to a larger sphere than the dimmer mass of the satellite which it partially clasps. Thus also, at considerable distances, the parallel flashes sent from a number of separate lamps and reflectors in a lighthouse encroach upon each other, and blend together to a single flash. The white-hot particles of carbon in a flame describe *lines* of light, because of their rapid upward motion. These lines are *widened* to the eye; and thus a far greater apparent *solidity* is imparted to the flame than in reality belongs to it.

189*a*. This augmentation of the true size of the optical image is called *Irradiation*.

Bodies seen within the Eye.

190. Almost every eye contains bodies more or less opaque distributed through its humours. The so-called *muscæ volitantes* are of this character; so are the black dots, snake-like lines, beads, and rings, which are strikingly visible in many eyes. Were the area of the pupil contracted to a point, such bodies might produce considerable annoyance; but because of the width of the pupil the shadows which these small bodies would otherwise cast upon the retina are practically obliterated, except when they are very close to the back of the eye.* It is only necessary to look at the firmament through a pinhole to give these shadows greater definition upon the retina.

191. The veins and arteries of the retina itself also cast their shadows upon its posterior surface; but the shaded spaces soon become so sensitive to light as to compensate for the defect of light falling upon them. Hence under ordinary circumstances the shadows are not seen. But if the shadows be transported to a less sensitive portion of the retina, the image of the vessels becomes distinctly visible.

192. The best mode of obtaining the transference of the shadow is to concentrate in a dark room, by means of a pocket lens of short focus, a small image of the sun or of the electric light upon the *white* of the eye. Care must be taken not to send the beam through the pupil. When the small lens is caused to move to and fro, the shadows are caused to travel over different portions of the retina, and a perfectly defined image of the veins and arteries is seen projected in the darkness in front of the eye.

193. Looking into a dark space, and moving a candle at the same time to and fro beside the eye, so that the rays enter the pupil very obliquely, the shadow of the retinal vessels is also obtained. In some eyes the suddenness and vigour with which the spectral image displays itself are extraordinary; others find it difficult to obtain the effect.

194. Finally, a delicate image of the vessels may be obtained by looking through a pinhole at the bright sky, and moving the aperture to and fro.

* See Notes 18 and 19.

The Stereoscope.

195. Look with one eye at the edge of the hand, so that the finger nearest the eye shall cover all the others. Then open the second eye; by it the other fingers will be seen foreshortened. *The images of the hand therefore within the two eyes are different.*

196. These two images are projected on the two retinæ; if by any means we could combine two drawings, executed on a flat surface, so as to produce within the two eyes two pictures similar to the two images of the solid hand, we should obtain the impression of *solidity.* This is done by the stereoscope.

197. The first form of this instrument was invented by Sir Charles Wheatstone. He took drawings of solid objects as seen by the two eyes, and looked at the images of these drawings in two plane mirrors. Each eye looked at the image which belonged to it, and the mirrors were so arranged that the images overlapped, thus appearing to come from the same object. By this combination of its two plane projections, the object sketched was caused to start forth as a solid.

198. In looking at and combining two such drawings, the eyes receive the same impression, and go through the same process as when they look at the real object. We see only one point of an object distinctly at a time. If the different points of an object be at different distances from the eyes, to see the near points distinctly requires a greater convergence of the axes of the eyes than to see the distant ones. Now, besides the identity of the retinal images of the stereoscopic drawings with those of the real object, the eyes, in order to cause the corresponding pairs of points of the two drawings to coalesce, have to go through the same variations of convergence that are necessary to see distinctly the various points of the actual object. Hence the impression of solidity produced by the combination of such drawings.

199. Measure the distance between *two pairs* of points, which when combined by the stereoscope present two *single points* at different distances from the eye. The distance between the one pair will be greater than that between the other pair. Different degrees of convergence are therefore necessary on the part of the eye to combine the two pairs of points. It is to be noted that the coalescence produced in the stereoscope at any particular moment is only *partial.* If one pair of corresponding points be seen singly, the others must appear double. This is also the case when an actual solid is looked at with both eyes; of those points of it which are at different distances from the eyes one only is seen singly at a time.

200. The impression of solidity may be produced in an exceedingly striking manner without any stereoscope at all. Most easily, thus :—Take two drawings—projections, as they are called—of the frustum of a cone; the one as it is seen by the right eye, the other as it is seen by the left. Holding them at some distance from the eyes, let the left-hand drawing be looked at by the right eye, and the

right-hand drawing by the left. The lines of vision of the two eyes here cross each other; and it is easy, after a few trials with a pencil-point placed in front of the eyes, to make two corresponding points of the drawings coincide. The moment they coincide, the combined drawings start forth as a single solid, suspended in the air at the place of intersection of the lines of vision. It depends upon the character of the drawings whether the inside of the frustum is seen, or the outside, whether its base or its top seems nearest to the eye. For this experiment the drawings are best made in simple outline, and they may be immensely larger than ordinary stereoscopic drawings.

Take notice that here also the different pairs of the corresponding points are at different distances apart. Two corresponding points, for example, of the top of the frustum will not be the same distance asunder as two points of the base.

201. Wheatstone's first instrument is called the Reflecting Stereoscope; but the methods of causing drawings to coalesce so as to produce stereoscopic effects are almost numberless. The instrument most used by the public is the Lenticular Stereoscope of Sir David Brewster. In it the two projections are combined by means of two half lenses with their edges turned inwards. The lenticular stereoscope also magnifies.*

202. It has been stated in note 198 that for the distinct vision of a near point a greater convergence of the lines of vision of the two eyes is necessary than for that of a distant one. By an instrument in which two rectangular prisms are employed,† the rays from two points may be caused to *cross* before they enter the eyes, the convergence being thus rendered greater for the distant point than for the near one. The consequence of this is, that the near point appears distant, and the distant point near. This is the principle of Wheatstone's *pseudoscope.* By this instrument convex surfaces are rendered concave, and concave surfaces convex. The inside of a hat or teacup may be thus converted into its outside, while a globe may be seen as a concave spherical surface.

Nature of Light; Physical Theory of Reflexion and Refraction.

It is now time to redeem to some extent the promise of our first note, that the 'something' which excites the sensation of light should be considered more closely subsequently.

203. Every sensation corresponds to a motion excited in our nerves. In the sense of touch, the nerves are moved by the contact of the body felt; in the sense of smell, they are stirred by the infinitesimal particles of the odorous body; in the sense of hearing, they are shaken by the vibrations of the air.

* Fuller and clearer information regarding the stereoscope will be found in the *Journal of the Photographic Society,* vol. iii. pp. 96, 116, and 167.
† See Note 150.

Theory of Emission.

204. Newton supposed light to consist of small particles shot out with inconceivable rapidity by luminous bodies, and fine enough to pass through the pores of transparent media. Crossing the humours of the eye, and striking the optic nerve behind the eye, these particles were supposed to excite vision.

205. This is the *Emission Theory* or *Corpuscular Theory* of Light.

206. Considering the enormous velocity of light, the particles, if they exist, must be inconceivably small; for if of any conceivable weight, they would infallibly destroy so delicate an organ as the eye. A bit of ordinary matter, one grain in weight, and moving with the velocity of light, would possess the momentum of a cannon-ball 150 lbs. weight, moving with a velocity of 1000 feet a second.

207. Millions of these light particles, supposing them to exist, concentrated by lenses and mirrors, have been shot against a balance suspended by a single spider's thread; this thread, though twisted 18,000 times, showed no tendency to untwist itself; it was therefore devoid of torsion. But no motion due to the impact of the particles was even in this case observed.*

208. If light consists of minute particles, they must be shot out with the same velocity by all celestial bodies. This seems exceedingly unlikely, when the different gravitating forces of such different masses are taken into account. By the attractions of such diverse masses, the particles would in all probability be pulled back with different degrees of force.

209. If, for example, a fixed star of the sun's density possessed 250 times the sun's diameter, its attraction, supposing light to be acted on like ordinary matter, would be sufficient to finally stop the particles of light issuing from it. Smaller masses would exert corresponding degrees of retardation; and hence the light emitted by different bodies would move with different velocities. That such is not the case—that light moves with the same velocity whatever be its source—renders it probable that it does *not* consist of particles thus darted forth.

But a more definite and formidable objection to the Emission Theory may be stated after we have made ourselves acquainted with the account it rendered of the phenomena of reflexion and refraction.

210. In direct reflexion, according to the emission theory, the light particles are first of all stopped in their course by a repellent force exerted by the reflecting body, and then driven in the contrary direction by the same force.

211. This repulsion is however *selective*. The reflecting substance singles out one portion of the group of particles composing a luminous beam and drives them back; but it attracts the remaining particles of the group and transmits them.

212. When a light particle approaches a refractive surface

* Bennett, *Phil. Trans.* 1792.

obliquely, if the particle be an attracted one, it is drawn towards the surface, as an ordinary projectile is drawn towards the earth. Refraction is thus accounted for. Like the projectile, too, the velocity of the light particle is *augmented* during its deflection; it enters the refracting medium with this increased velocity, and once within the medium, the attractions before and behind the particle neutralising each other, the increased velocity is maintained.

213. Thus, it is an unavoidable consequence of the theory of Newton, that the bending of a ray of light towards the perpendicular is accompanied by an augmentation of velocity—that light in water moves more rapidly than in air, in glass more rapidly than in water, in diamond more rapidly than in glass. In short, that the higher the refractive index, the greater the velocity of the light.

214. A decisive test of the emission theory was thus suggested, and under that test the theory has broken down. For it has been demonstrated, by the most rigid experiments, that the velocity of light *diminishes* as the index of refraction increases. The theory, however, had yielded to the assaults made upon it long before this particular experiment was made.

Theory of Undulation.

215. The Emission Theory was first opposed by the celebrated astronomer Huygens and the no less celebrated mathematician Euler, both of whom held that light, like sound, was a product of *wave motion.* Laplace, Malus, Biot, and Brewster supported Newton, and the emission theory maintained its ground until it was finally overthrown by the labours of Thomas Young* and Augustin Fresnel.

216. These two eminent philosophers, while adducing whole classes of facts inexplicable by the emission theory, succeeded in establishing the most complete parallelism between optical phenomena and those of wave motion. The justification of a theory consists in its exclusive competence to account for phenomena. On such a basis the *Wave Theory,* or the *Undulatory Theory* of light, now rests, and every day's experience only makes its foundations more secure. This theory must for the future occupy much of our attention.

---

* Dr. Young was appointed Professor of Natural Philosophy in the Royal Institution, August 3, 1801. From a marble slab in the village church of Farnborough, near Bromley, Kent, I copied, on the 11th of April, the following inscription :—

'Near this place are deposited the remains of THOMAS YOUNG, M.D., Fellow and Foreign Secretary of the Royal Society, Member of the National Institute of France. A man alike eminent in almost every department of human learning, whose many discoveries enlarged the bounds of Natural Science, and who first penetrated the obscurity which had veiled for ages the Hieroglyphics of Egypt.

'Endeared to his friends by his domestic virtues, Honoured by the world for his unrivalled acquirements, He died in the hope of the resurrection of the just.

'Born at Milverton, in Somersetshire, June 13th, 1773.

'Died in Park Square, London, May 29th, 1829,

'In the 56th year of his age.'

217. In the case of sound, the velocity depends upon the relation of elasticity to density in the body which transmits the sound. The greater the elasticity the greater is the velocity, and the less the density the greater is the velocity. To account for the enormous velocity of propagation in the case of light, the substance which transmits it is assumed to be of both extreme elasticity and of extreme tenuity. This substance is called the *Luminiferous ether.*

218. It fills space; it surrounds the atoms of bodies; it extends, without solution of continuity, through the humours of the eye. The molecules of luminous bodies are in a state of vibration. The vibrations are taken up by the ether, and transmitted through it in waves. These waves impinging on the retina excite the sensation of light.

219. In the case of sound, the air-particles oscillate to and fro in the direction *in* which the sound is transmitted; in the case of light, the ether particles oscillate to and fro *across* the direction in which the light is propagated. In scientific language the vibrations of sound are *longitudinal*, while the vibrations of light are *transversal*. In fact, the mechanical properties of the ether are rather those of a solid than of an air.

220. The *intensity* of the light depends on the distance to which the ether particles move to and fro. This distance is called the *amplitude* of the vibration. The intensity of light is proportional to the *square* of the amplitude; it is also proportional to the square of the maximum velocity of the vibrating particle.

221. The amplitude of the vibrations diminishes simply as the distance increases; consequently the intensity, which is expressed by the square of the amplitude, must diminish inversely as the square of the distance. This, in the language of the wave theory, is the law of inverse squares.

222. The reflexion of ether waves obeys the law established in the case of light. The angle of incidence is demonstrably equal to the angle of reflexion.

223. To account for refraction, let us for the sake of simplicity take a portion of a circular wave emitted by the sun or some other distant body. A short portion of such a wave would be *straight.* Suppose it to impinge from air upon a plate of glass, the wave being in the first instance *parallel* to the surface of the glass. Such a wave would go through the glass without change of direction.

224. But as the velocity in glass is less than the velocity in air, the wave would be *retarded* on passing into the denser medium.

225. But suppose the wave, before impact, to be *oblique* to the surface of the glass; that end of the wave which first reaches the glass will be first retarded, the other portions being held back in succession. This retardation of one end of the wave causes it to swing round; so that when the wave has fully entered the glass its *course is* oblique to its first direction. It is *refracted.*

226. If the glass into which the wave enters be a plate with parallel surfaces, the portion of the wave which reached the upper surface *first*, and was first retarded, will also reach its under surface first, and escape earliest from the retarding medium. This produces a second swinging round of the wave, by which its original direction is restored. In this simple way the Wave Theory accounts for Refraction.

227. The convergence or divergence of beams of light by lenses is immediately deduced from the fact that the different points of the ether wave reach the lens, and are retarded by the lens in succession.

228. The density of the ether is greater in liquids and solids than in gases, and greater in gases than in vacuo. A compressing force seems to be exerted on the ether by the molecules of these bodies. Now if the elasticity of the ether increased in the same proportion as its density, the one would neutralise the other, and we should have no retardation of the velocity of light. The diminished velocity in highly refracting bodies is accounted for by assuming that in such bodies the elasticity *in relation to the density* is less than in vacuo. The observed phenomena immediately flow from this assumption.

229. The case is precisely similar to that of sound in a gas or vapour which does not obey the law of Mariotte. The elasticity of such a gas or vapour, when compressed, increases less rapidly than the density; hence the diminished velocity of the sound.

230. But we are able to give a more distinct statement as to the influence which a refracting body has upon the velocity of light. Regard the lines *o m* and *np* in Fig. 2, Note 113. These two lines *represent the velocities of light* in the two media there considered; or, expressed more generally, the sine of the angle of incidence represents the velocity of light in the first medium, while the sine of the angle of refraction represents the velocity in the second. *The index of refraction then is nothing else than the ratio of the two velocities.* Thus in the case of water where the index of refraction is $\frac{4}{3}$ the velocity in air is to its velocity in water as 4 is to 3. In glass also, where the index of refraction is $\frac{3}{2}$ the velocity in air is to the velocity in glass as 3 is to 2. In other words the velocity of light in air is $1\frac{1}{3}$ times its velocity in water, and $1\frac{1}{2}$ times its velocity in glass. The velocity of light in air is about $2\frac{1}{3}$ times its velocity in diamond, and nearly three times its velocity in chromate of lead, the most powerfully refracting substance hitherto discovered. Strictly speaking, the index of refraction refers to the passage of a ray of light, not from *air*, but from a vacuum* into the refracting body. Dividing the velocity of light in vacuo by its velocity in the refracting substance, the quotient is the index of refraction of that substance.

231. In the wave theory, the rays of light are perpendiculars to the waves of ether. Unlike the *wave*, the *ray* has no material existence; it is merely a direction.

* That is to say, a vacuum save as regards the ether itself.

D

Prisms.

232. It has been stated in note 129, that in the case of a plate of glass *with parallel surfaces*, the direction possessed by an oblique ray, prior to its meeting the glass, is restored when it quits the glass. This is not the case if the two surfaces at which the ray enters and emerges be not parallel.

233. When the ray passes through a wedge-shaped transparent substance, in a direction perpendicular to the edge of the wedge, it is *permanently* refracted. A body of this shape is called a *prism* in optics, and the angle enclosed by the two oblique sides of the wedge is called the *refracting angle*.

234. The larger the refracting angle the greater is the deflection of the ray from its original direction. But with the self-same prism the amount of the refraction varies with the direction pursued by the ray through the prism.

235. When that direction is such that the portion of the ray within the prism makes equal angles with the two sides of the prism, or what is the same, with the ray before it reaches the prism and after it has quitted it, then the total refraction is a *minimum*. This is capable both of mathematical and experimental proof; and on this result is based a method of determining the index of refraction.

236. The final direction of a refracted ray being unaltered by its passage through glass plates with parallel surfaces, we may employ hollow vessels composed of such plates and filled with liquids, thus obtaining liquid prisms.

Prismatic Analysis of Light: Dispersion.

237. Newton first unravelled the solar light, proving it to be composed of an infinite number of rays of different degrees of refrangibility; when such light is sent through a prism, its constituent rays are drawn asunder. This act of drawing apart is called *dispersion*.

238. The waves of ether generated by luminous bodies are not all of the same length; some are longer than others. In refracting substances the short waves are *more retarded* than the long ones; hence the short waves are more *refracted* than the long ones. This is the cause of dispersion.

239. The luminous image formed when a beam of white light is thus decomposed by a prism is called a *spectrum*. If the light employed be that of the sun, the image is called the solar spectrum.

240. The solar spectrum consists of a series of vivid colours, which, when reblended, produce the original white light. Commencing with that which is least refracted, we have the following *order of colours* in the solar spectrum :—Red, Orange, Yellow, *Green, Blue,* Indigo, Violet.

241. *The Colour of Light is determined solely by its Wave-length.*
—The ether waves gradually diminish in length from the red to the
violet. The length of a wave of red light is about $\frac{1}{39000}$ of an inch:
that of a wave of violet light is about $\frac{1}{57500}$th of an inch. The
waves which produce the other colours of the spectrum lie between
these extremes.

242. The velocity of light being 192,000 miles in a second, if
we multiply this number by 39,000 we obtain the number of waves
of red light in 192,000 miles; the product is 474,439,680,000,000.
All these waves enter the eye in a second. In the same interval
699,000,000,000,000 waves of violet light enter the eye. At this
prodigious rate is the retina hit by the waves of light.

243. Colour, in fact, is to light what *pitch* is to sound. The pitch
of a note depends solely on the number of aerial waves which strike
the ear in a second. The colour of light depends on the number of
ethereal waves which strike the eye in a second. Thus the sensation
of red is produced by imparting to the optic nerve four hundred and
seventy-four millions of millions of impulses per second, while the
sensation of violet is produced by imparting to the nerve six hundred
and ninety-nine millions of millions of impulses per second. In the
Emission Theory numbers not less immense occur, ‘ nor is there any
mode of conceiving the subject which does not call upon us to admit
the exertion of mechanical forces which may well be termed infinite.’ *

Invisible rays: Calorescence and Fluorescence.

244. The spectrum extends in both directions beyond its visible
limits. Beyond the visible red we have rays which possess a high
heating power, though incompetent to excite vision; beyond the
violet we have a vast body of rays which, though feeble as regards
heat, and powerless as regards light, are of the highest importance
because of their capacity to produce chemical action.

245. In the case of the electric light, the energy of the non-
luminous calorific rays emitted by the carbon points is about eight
times that of all the other rays taken together. The dark calorific
rays of the sun also probably exceed many times in power the luminous solar rays. It is possible to sift the solar or the electric beam
so as to intercept the luminous rays, while the non-luminous rays
are allowed free transmission.

246. In this way perfectly dark foci may be obtained where combustible bodies may be burned, non-refractory metals fused, and
refractory ones raised to the temperature of whiteness. The non-
luminous calorific rays may be thus transformed into luminous ones,
which yield all the colours of the spectrum. This passage, by the
intervention of a refractory body, from the non-luminous to the luminous state, is called *Calorescence.*

* Sir John Herschel.

247. So also as regards the ultra-violet rays; when they are permitted to fall upon certain substances—the disulphate of quinine for example—they render the substance luminous; *invisible rays are thereby rendered visible.* The change here receives the name of *Fluorescence.*

248. In calorescence the atoms of the refractory body are caused to vibrate more rapidly than the waves which fall upon them; the periods of the waves are quickened by their impact on the atoms. The refrangibility of the rays is, in fact, *exalted.* In fluorescence, on the contrary, the impact of the waves throws the molecules of the fluorescent body into vibrations of slower periods than those of the incident waves; the refrangibility of the rays is in fact *lowered.* Thus by exalting the refrangibility of the ultra-red, and by lowering the refrangibility of the ultra-violet rays, both classes of rays are rendered capable of exciting vision.

249. Though the term is by no means faultless, those rays, both ultra-red and ultra-violet, which are incompetent to excite vision, are called *invisible rays.* In strictness we cannot speak of rays being either visible or invisible; it is not the rays themselves but the objects they illuminate that become visible. ' *Space,* though traversed by the rays from all suns and all stars, is itself unseen. Not even the ether which fills space, and whose motions are the light of the world, is itself visible.' *

Doctrine of Visual Periods.

250. A string tuned to a certain note resounds when that note is sounded. If you sing into an open piano, the string whose note is in unison with your voice will be thrown into sonorous vibration. If there be discord between the note and the string, there is no resonance, however powerful the note may be. A particular church-pane is sometimes broken by a particular organ-peal, through the coincidence of its period of vibration with that of the organ.

251. In this way it is conceivable that a feeble note, through the coincidence of its periods of vibration with those of a sounding body, may produce effects which a powerful note, because of its non-coincidence, is unable to produce.

252. This, which is a known phenomenon of sound, helps us to a conception of the deportment of the retina towards light. The retina, or rather the brain in which its fibres end, is, as it were, attuned to a certain range of vibrations, and it is dead to all vibrations which lie without that range, however powerful they may be.

253. The quantity of wave motion sent to the eye at night, by a candle a mile distant, suffices to render the candle visible. Employing the powerful ultra-red rays of the sun, or of the electric light,

* ' Proceedings of the Royal Institution,' vol. v., p. 456.

it is demonstrable that ethereal waves possessing many millions of times the mechanical energy of those which produce the candle's light, may be caused to impinge upon the retina without exciting any sensation whatever. *The periods of succession* of the waves, rather than their *strength*, are here influential.

254. When in music two notes are separated from each other by an octave, the higher note vibrates with twice the rapidity of the lower. In Note 241 the lengths of the wave of red light and of violet light are set down as $\frac{1}{39000}$ of an inch and $\frac{1}{57000}$ of an inch respectively; but these numbers refer to the *mean* red and the *mean* violet. The waves of the *extreme* violet are about half the length of those of the extreme red, and they strike the retina with double the rapidity of the red. While, therefore, the *musical scale*, or the range of the ear, is known to embrace nearly eleven octaves, the *optical scale*, or range of the eye, is comprised within a single octave.

Doctrine of Colours.

255. Natural bodies possess the power of extinguishing, or, as it is called, *absorbing* the light that enters them. This power of absorption is *selective*, and hence, for the most part, arise the phenomena of *colour*.

256. When the light which enters a body is *wholly* absorbed the body is black; a body which absorbs all the waves equally, but not totally, is grey; while a body which absorbs the various waves unequally is *coloured*. Colour is due to the extinction of certain constituents of the white light within the body, the remaining constituents which return to the eye imparting to the body its colour.

257. It is to be borne in mind that bodies of all colours, illuminated by white light, reflect white light *from their exterior surfaces*. It is the light which has plunged to a certain depth within the body, which has been *sifted* there by elective absorption, and then discharged from the body by interior reflexion that, in general, gives the body its colour.

258. A pure red glass interposed in the path of a beam decomposed by a prism, either before or after the act of decomposition, cuts off all the colours of the spectrum except the red. A glass of any other pure colour similarly interposed would cut off all the spectrum except that particular portion which gives the glass its colour. It is, however, extremely difficult, if not impossible, to obtain pure pigments of any kind. Thus a yellow glass not only allows the yellow light of the spectrum to pass, but also a portion of the adjacent green and orange; while a blue glass not only allows the blue to pass, but also a portion of the adjacent green and indigo.

259. Hence, if a beam of white light be caused to pass through a yellow glass and a blue glass at the same time, the only transmissible colour *common to both* is green. This explains why blue and yellow

powders, when mixed together, produce green. The white light plunges into the powder to a certain depth, and is discharged by internal reflexion, *minus* its yellow and its blue. The green alone remains.

260. The effect is quite different when, instead of mixing blue and yellow *pigments*, we mix blue and yellow *lights* together. Here the mixture is a pure *white*. Blue and yellow are complementary colours.

261. Any two colours whose mixture produces white are called *complementary colours*. In the spectrum we have the following pairs of such colours :—

> Red and greenish Blue.
> Orange and cyanogen Blue.
> Yellow and indigo Blue.
> Greenish yellow and Violet.

262. A body placed in a light which it is incompetent to transmit appears black, however intense may be the illumination. Thus, a stick of red sealing-wax, placed in the vivid green of the spectrum, is perfectly black. A bright red solution similarly placed cannot be distinguished from black ink ; and red cloth, on which the spectrum is permitted to fall, shows its colour vividly where the red light falls upon it, but appears black beyond this position.

263. We have thus far dealt with the *analysis* of white light. In reblending the constituent colours, so as to produce the original, we illustrate, by *synthesis*, the composition of white light.

264. Let the beam analyzed be a rectangular slice of light. By means of a cylindrical lens we can recombine the colours, and produce by their mixture the original white. It is also possible, by the combination of the colours of its spectrum, to build up a perfect image of the source of light. The persistence of impressions on the retina also offers a ready means of blending colours.

Chromatic Aberration.   Achromatism.

265. Owing to the different refrangibility of the different rays of the spectrum, it is impossible by a single spherical lens to bring them all to a focus at the same point. The blue rays, for example, being more refracted than the red will intersect sooner than the red.

266. Hence, when a divergent cone of white light is rendered convergent by a lens, the convergent beam, as far as the point of intersection of the rays, will be surrounded by a sheath of red ; while beyond the focus the divergent cone will be surrounded by a sheath of blue. Hence, when the refracted rays fall upon a screen placed between the lens and the focus of blue rays, a white circle with a red border is obtained, while if the screen be placed beyond the focus of *red rays* the white circle will have a blue border. It is impossible *to produce a colourless* image in these positions of the screen.

267. This lack of power on the part of a lens to bring its differently

coloured constituents to a common focus, is called the *Chromatic aberration* of the lens.

268. Newton considered it impossible to get rid of chromatic aberration; for he supposed the dispersion of a prism or lens to be proportional to its refraction, and that if you destroyed the one you destroyed the other. This, however, was an error.

269. For two prisms producing the same mean refraction may produce very different degrees of dispersion. By diminishing the angle of the more highly dispersive prism we can make its dispersion sensibly equal to that of the feebly dispersive one; and we can neutralize the colours of both prisms by placing them in opposition to each other, without neutralizing the refraction.

270. When, for example, a prism of water is opposed to a prism of flint-glass, after the dispersion of the water, which is small, has been destroyed, the beam is still refracted. If a prism of *crown-glass* be substituted for the prism of water, substantially the same effect is produced. The flint-glass is competent to neutralize the dispersion of the crown *before* it neutralizes the refraction.

271. What is here said of prisms applies equally to lenses. A convex crown-glass lens, opposed to a concave flint-glass lens, may have its dispersion destroyed, and still images may be formed by the combination of the two lenses, because of the *residual* refraction.

272. A combination of lenses wherein colour is destroyed while a certain amount of refraction is preserved, is called an achromatic combination, or more briefly an *achromatic lens*.

273. The human eye is not achromatic. It suffers from chromatic aberration as well as from spherical aberration.

Subjective Colours.

274. By the action of light the optic nerve is rendered less sensitive. When we pass from bright daylight into a moderately lighted room, the room appears dark.

275. This is also true of individual colours; when light of any particular colour falls upon the eye the optic nerve is rendered less sensitive to that colour. It is, in fact, partially blinded to its perception.

276. If the eyes be steadily fixed upon a red wafer placed on white paper, after a little time the wafer will be surrounded by a greenish rim, and if the wafer be moved away, the place on which it rested will appear green.

277. This is thus explained:—the eye by looking at the wafer has its sensibility to red light diminished; hence, when the wafer is removed, the white light falling upon the spot of the retina on which the image of the wafer rested, will have its red constituent virtually removed, and will therefore appear of the complementary colour. The first rim of green light observed is due to the extension of the red

light of the wafer a little beyond its geometrical image on the retina, in consequence of the spherical aberration of the eye.

278. Coloured shadows are reducible to the same cause. Let a strong red light, for example, fall upon a white screen. A body interposed between the light and the screen will cast a shadow, and if this shadow be moderately illuminated by a second white light it will appear green. If the original light be blue, the shadow will appear yellow; if the original light be green, the shadow will appear red. The reason is, that the eye in the first instance is partially blinded to the perception of the colour cast upon the screen; hence the white light, which reaches the eye from the shadow, will have that colour partially withdrawn, and the shadow will appear of the complementary colour.

279. Colours of this kind are called *subjective colours*; they depend upon the condition of the eye, and do not express external facts of colour.

Spectrum Analysis.

280. Metals and their compounds impart to flames peculiar colours, which are characteristic of the metals. Thus the almost lightless flame of a Bunsen's burner is rendered a brilliant yellow by the metal sodium, or by any volatilizible compound of that metal, such as chloride of sodium or common salt. The flame is rendered green by copper, purple by zinc, and red by strontian.

281. These colours are due to the *vapours* of the metals which are liberated in the flame.

282. When such incandescent metallic vapours are examined by the prism, it is found that instead of emitting rays which form a *continuous* spectrum, one colour passing gradually into another, they emit distinct groups of rays of definite, but different refrangibilities. The spectrum corresponding to these rays is a series of coloured bands, separated from each other by intervals of darkness. Such bands are characteristic of luminous gases of all kinds.

283. Thus the spectrum of incandescent sodium-vapour consists of a brilliant band on the confines of the orange and yellow; and the vapour is incompetent to shed forth any of the other light of the spectrum. When this band is more accurately analyzed it resolves itself into two distinct bands; greater delicacy of analysis resolves it into *a group* of bands with fine dark intervals between them. The spectrum of copper-vapour is signalized by a series of green bands, while the incandescent vapour of zinc produces brilliant bands of blue and red.

284. The light of the bands produced by metallic vapours is very *intense,* the whole of the light being concentrated into a few narrow *strips, and escaping* in a great measure the dilution due to dispersion.

285. These coloured bands are perfectly characteristic of the vapour; from their position and number the substance that produces them can be unerringly inferred.

286. If two or more metals be introduced into the flame at the same time, prismatic analysis reveals the bands of each metal as if the others were not there. This is also true when a mineral containing several metals is introduced into the flame. The constituent metals of the mineral will give each its characteristic bands.

287. Hence, having made ourselves acquainted with the bands produced by all known metals, if entirely new bands show themselves, it is a proof that an entirely new metal is present in the flame. It is thus that Bunsen and Kirchhoff, the founders of spectrum analysis, discovered Rubidium and Cæsium; and that Thallium, with its superb green band, was discovered by Mr. Crookes.

288. The *permanent gases* when heated to a sufficient temperature, as they may be by the electric discharge, also exhibit characteristic bands in their spectra. By these bands they may be recognized, even at stellar distances.

289. The action of light upon the eye is a test of unrivalled delicacy. In *spectrum analysis* this action is brought specially into play; hence the power of this method of analysis.*

Further Definition of Radiation and Absorption.

290. The terms ray, radiation, and absorption, were employed long prior to the views now entertained regarding the nature of light. It is necessary more clearly to understand the meaning attached by the undulatory theory to those terms.

291. And to complete our knowledge it is necessary to know that all bodies, whether luminous or non-luminous, are *radiants*; if they do not radiate light they radiate heat.

292. It is also necessary to know that luminous rays are also heat rays; that the self-same waves of ether falling on a thermometer produce the effects of heat; and impinging upon the retina produce the sensation of light. The rays of greatest heat however, as already explained, lie entirely without the visible spectrum.

293. The radiation both of light and heat consists in the *communication* of motion from the vibrating atoms of bodies to the ether which surrounds them. The absorption of heat consists in the *acceptance* of motion, on the part of the atoms of a body, from ether which

---

* Many persons are incompetent to distinguish one colour of the spectrum from another; red and green, for example, are often confounded. Dalton, the celebrated founder of the Atomic Theory, could only distinguish by their form ripe red cherries from the green leaves of the tree. This point is now attended to in the choice of engine-drivers, who have to distinguish one coloured signal from another. *The defect is called* colour-blindness, *and sometimes* Daltonism.

has been already agitated by a source of light or heat. In radiation, then, motion is yielded to the ether; in absorption, motion is received from the ether.

294. When a ray of light or of heat passes through a body without loss; in other words, when the waves are transmitted *through the ether* which surrounds the atoms of the body, without sensibly imparting motion to the atoms themselves, the body is *transparent*. If motion be in any degree transferred from the ether to the atoms, in that degree is the body *opaque*.

295. If either light or radiant heat be absorbed, the absorbing body is *warmed*; if no absorption takes place, the light or radiant heat, whatever its intensity may be, passes through the body without affecting its temperature.

296. Thus in the dark foci referred to in Note 246, or in the focus of the most powerful burning mirror which concentrates the beams of the sun, the *air* might be of a freezing temperature, because the absorption of the heat by the air is insensible. A plate of clear rock-salt, moreover, placed at the focus, is scarcely sensibly heated, the absorption being small; while a plate of glass is shivered, and a plate of blackened platinum raised to a white heat, or even fused, because of their powers of absorption.

297. It is here worth remarking that calculations of the temperatures of comets, founded on their distances from the sun, may be, and probably are, entirely fallacious. The comet, even when nearest to the sun, might be intensely cold. It might carry with it round its perihelion the chill of the most distant regions of space. If transparent to the solar rays it would be unaffected by the solar heat, as long as that heat maintained *the radiant form.*

The pure Spectrum : Fraunhofer's Lines.

298. When a beam of white light issuing from a slit is decomposed, the spectrum really consists of a series of coloured images of the slit placed side by side. If the slit be wide, these images *overlap*; but in a *pure* spectrum the colours must not overlap each other.

299. A pure spectrum is obtained by making the slit through which the decomposed beam passes very narrow, and by sending the beam through several prisms in succession, thus augmenting the dispersion.

300. When the light of the sun is thus treated, the solar spectrum is found to be not perfectly continuous; across it are drawn innumerable dark lines, the rays corresponding to which are absent. Dr. Wollaston was the first to observe some of these lines. They were afterwards studied with supreme skill by Fraunhofer, who lettered *them and made* accurate maps of them, and from him they have been *called Fraunhofer's lines.*

Reciprocity of Radiation and Absorption.

301. To account for the missing rays of the lines of Fraunhofer was long an enigma with philosophers. By the genius of Kirchhoff the enigma was solved. Its solution carried with it a new theory of the constitution of the sun, and a demonstration of a method which enables us to determine the chemical composition of the sun, the stars, and the nebulæ. The application of Kirchhoff's principles by Messrs. Huggins, Miller, Secchi, Janssen, and Lockyer has been of especial interest and importance.

302. Kirchhoff's explanation of the lines of Fraunhofer is based upon the principle that every body is specially opaque to such rays as it can itself emit when rendered incandescent.

303. Thus the radiation from a carbonic oxide flame, which contains carbonic acid at a high temperature, is intercepted in an astonishing degree by carbonic acid. If the rays from a sodium flame be sent through a second sodium flame, they will be stopped with particular energy by the second flame. The rays from incandescent thallium vapour are intercepted by thallium vapour, those from lithium vapour by lithium vapour, and so of the other metals.

304. In the language of the undulatory theory, waves of ether are absorbed with special energy—their motion is taken up with special facility—by atoms whose periods of vibration synchronise with the periods of the waves. This is another way of stating that a body absorbs with special energy the rays which it can itself emit.

305. If a beam of white light be sent through the intensely yellow flame of sodium vapour, the yellow constituent of the beam is intercepted by the flame, while rays of other refrangibilities are allowed free transmission.

306. Hence, when the spectrum of the electric light is thrown upon a white screen, the introduction of a sodium flame into the path of the rays cuts off the yellow component of the light, and the spectrum is furrowed by a dark band in place of the yellow.

307. Introducing other flames in the same manner in the path of the beam, if the quantity of metallic vapour in the flame be sufficient, each flame will cut out its own bands. And if the flame through which the light passes contain the vapours of several metals, we shall have the dark characteristic bands of all of them upon the screen.

308. Expanding in idea our electric light until it forms a globe equal to the sun in size, and wrapping round this incandescent globe an atmosphere of flame, that atmosphere would cut off those rays of the globe which it can itself emit, the interception of the rays being declared by dark lines in the spectrum.

309. We thus arrive at a complete explanation of the lines of Fraunhofer, and a new theory of the constitution of the sun. The orb *consists of a solid or molten nucleus, in a condition of intense*

incandescence, but it is surrounded by a gaseous photosphere containing vapours which absorb those rays of the nucleus which they themselves emit. The lines of Fraunhofer are thus produced.

310. The lines of Fraunhofer are narrow bands of *partial* darkness; they are really illuminated by the light of the gaseous envelope of the sun. But this is so feeble in comparison with the light of the nucleus intercepted by the envelope, that the bands appear dark in comparison with the adjacent brilliance.

311. Were the central nucleus abolished, the bands of Fraunhofer *on a perfectly dark ground*, would be transformed into a series of bright bands. These would resemble the spectra obtained from a flame charged with metallic vapours. They would constitute the spectrum of the solar atmosphere.

312. It is not necessary that the photosphere should be composed of *pure vapour*. Doubtless it contains vast masses of incandescent cloudy matter, composed of white hot molten particles. These intensely luminous white hot clouds may be the main origin of the light which the earth receives from the sun, and with them the true vapour of the photosphere may be more or less confusedly mingled. But the vapour which produces the lines of Fraunhofer must exist *outside* the clouds, as assumed by Kirchhoff.

Solar Chemistry.

313. From the dark bands of the spectrum we can determine what substances enter into the composition of the solar atmosphere.

314. One example will illustrate the possibility of this. Let the light from the sun and the light from incandescent sodium vapour pass side by side through the same slit, and be decomposed by the same prism. The solar light will produce its spectrum, and the sodium light its yellow band. This yellow band will coincide exactly in position with a characteristic dark band of the solar spectrum, which Fraunhofer distinguishes by the letter D.

315. Were the solar nucleus absent, and did the vaporous photosphere alone emit light, the dark line D would be a bright one. Its character and position prove it to be the light emitted by sodium. This metal, therefore, is contained in the atmosphere of the sun.[*]

316. The result is still more convincing when a metal which gives a numerous series of bright bands finds each of its bands exactly coincident with a dark band of the solar spectrum. By this method Kirchhoff, to whom we owe, in all its completeness, this splendid generalization, established the existence of iron, calcium,

---

[*] By reference to note 283 it will be seen that the sodium line is resolved by delicate analysis into a group of lines. The Fraunhofer dark band D is similarly resolved. It ought to be mentioned that both Mr. Talbot and Sir John Herschel clearly foresaw the possibility of employing spectrum analysis in detecting *minute* traces of bodies.

magnesium, sodium, chromium, and other metals in the solar atmosphere; and Mr. Huggins has extended the application of the method to the light of the planets, fixed stars, and nebulæ.*

Planetary Chemistry.

317. The light reflected from the moon and planets is solar light; and, if unaffected by the planet's atmosphere, the spectrum of the planet would show the same lines as the solar spectrum.

318. The light of the moon shows no other lines. There is no evidence of an atmosphere round the moon.

319. The lines in the spectrum of Jupiter indicate a powerful absorption by the atmosphere of this planet. The atmosphere of Jupiter contains some of the gases or vapours present in the earth's atmosphere. Feeble lines, some of them identical with those of Jupiter, occur in the spectrum of Saturn.

320. The lines characterizing the atmospheres of Jupiter and Saturn are not present in the spectrum of Mars. The blue portion of the spectrum is mainly the seat of absorption; and this, by giving predominance to the red rays, may be the cause of the red colour of Mars.

321. All the stronger lines of the solar spectrum are found in the spectrum of Venus, but no additional lines.

Stellar Chemistry.

322. The atmosphere of the star Aldebaran contains hydrogen, sodium, magnesium, calcium, iron, bismuth, tellurium, antimony, mercury. The atmosphere of the star Alpha in Orion contains sodium, magnesium, calcium, iron, and bismuth.

323. No star sufficiently bright to give a spectrum has been observed to be without lines. Star differs from star only in the grouping and arrangement of the numerous fine lines by which their spectra are crossed.

324. The dark absorption lines are strongest in the spectra of yellow and red stars. In white stars the lines, though equally numerous, are very poor and faint.

325. A comparison of the spectra of stars of different colours suggests that the colours of the stars may be due to the action of their atmospheres. Those constituents of the white light of the star on which the lines of absorption fall thickest are subdued, the star being tinted by the residual colour.

Father Secchi, of Rome, has studied the light of many hundreds of stars, and has divided them into four classes.

---

* Professor Stokes foresaw the possible application of spectrum analysis to solar chemistry.

Nebular Chemistry.

326. Some nebulæ give spectra of bright bands, others give continuous spectra. The light from the former emanates from intensely heated matter existing *in a state of gas.* This may in part account for the weakness of the light of these nebulæ.

327. It is probable that two of the constituents of the gaseous nebulæ are hydrogen and nitrogen.

The Red Prominences and Envelope of the Sun.

328. Astronomers had observed during total eclipses of the sun vast red prominences extending from the solar limb many thousand miles into space. The intense illumination of the circum-solar region of our atmosphere masks, under ordinary circumstances, the red prominences. They are quenched, as it were, by excess of light.

329. But when, by the intervention of the dark body of the moon, this light is cut off, the prominences are distinctly seen.

330. It was proved by Mr. De la Rue and others that the red matter of the prominences was wrapped round a large portion of the sun's surface. According to the observations of Mr. Lockyer, the red matter forms a *complete envelope* round the sun.

331. Examined by the spectroscope the matter of the prominences shows itself to be, for the most part, incandescent hydrogen. With it are mixed the vapours of sodium and magnesium.

332. Mr. Janssen in India, and Mr. Lockyer subsequently, but independently, in England proved that the bright bands of the prominences might be seen without the aid of a total eclipse. The explanation of this discovery is glanced at in Note 284, where the intensity of the bright bands of incandescent gases was referred to the practical absence of dispersion.

333. By sending the light, which under ordinary circumstances masks the hydrogen bands, through a sufficient number of prisms it may be dispersed, and thereby enfeebled in any required degree. When sufficiently enfeebled the undispersed light of the incandescent hydrogen dominates over that of the continuous spectrum. By going completely round the periphery of the sun Mr. Lockyer found this hydrogen atmosphere everywhere present, its depth, generally about 5,000 miles, being indicated by the length of its characteristic bright lines. Where the hydrogen ocean is shallow the bright bands are short, where the prominences rise like vast waves above the level of the ocean the bright lines are long. The prominences sometimes reach a height of 70,000 miles.

The Rainbow.

334. A beam of solar light, falling obliquely on the surface of a *rain-drop, is* refracted on entering the drop; it is in part reflected at the back of the drop, and on emerging from the drop it is again refracted.

335. By these two refractions on entrance and on emergence the beam of light is decomposed, and it quits the drop resolved into its coloured constituents. It is received by the eye of an observer who faces the drop and turns his back to the sun.

336. In general the solar rays, when they quit the drop, are *divergent*, and therefore produce but a feeble effect upon the eye. But at *one* particular angle the rays, after having been twice refracted and once reflected, issue from the drop almost perfectly parallel. They thus preserve their intensity like rays reflected from a parabolic mirror, and produce a corresponding effect upon the eye. The angle at which this parallelism is established varies with the refrangibility of the light.

337. Draw a line from the sun to the observer's eye and prolong this line beyond the observer. Conceive another line drawn from the eye enclosing an angle of 42° 30′ with the line drawn to the sun. The rain-drop struck by this second line will send to the eye a parallel beam of *red light*. Every other drop similarly situated, that is to say, every drop at an angular distance of 42° 30′ from the line drawn to the sun will do the same. We thus obtain a *circular band* of red light, forming part of the base of a cone, of which the eye of the observer is the apex. Because of the angular magnitude of the sun the width of this band will be half a degree.

338. From the eye of the observer conceive another line to be drawn enclosing an angle of 40° 30′ with the line drawn to the sun. A drop struck by this line will send along the line an almost perfectly parallel beam of *violet* light to the eye. All drops at the same angular distance will do the same, and we shall obtain a band of violet light of the same width as the red. These two bands constitute the limiting colours of the rainbow, and between them the bands corresponding to the other colours lie.

339. The rainbow is in fact a spectrum, in which the rain-drops play the part of prisms. The width of the bow from red to violet is about two degrees. The size of the arc visible at any time manifestly depends upon the position of the sun. The bow is grandest when it is formed by the rising or the setting sun. An entire semi-circle is then seen by an observer on a plain, while from a mountain-top a still greater arc is visible.

340. The angular distances and the order of colours here given correspond to the *primary bow*, but in addition to this we usually see a *secondary bow* of weaker hues, and in which the order of the colours is that of the primary *inverted*. In the primary the red band forms the convex surface of the arch; it is the largest band; in the secondary the violet band is outside, the red forming the concavity of the bow.

341. The secondary bow is produced by rays which have undergone *two* reflexions within the drop, as well as two refractions at its

surface. It is this double internal reflexion that weakens the colour. In the primary bow the incident rays strike the upper hemisphere of the drop, and emerge from the lower one; in the secondary bow the incident rays strike the lower hemisphere of the drop, emerge from the upper one, and then cross the incident rays to reach the eye of the observer. The secondary bow is $3\frac{1}{2}$ degrees wide, and it is $7\frac{1}{2}$ degrees higher than the primary. From the space between the two bows part of the light reflected from the *anterior surfaces* of the raindrops reaches the eye; but no light whatever that *enters* the raindrops in this space is reflected to the eye. Hence this region of the falling shower is darkest.

Interference of Light.

342. In wave motion we must clearly distinguish the motion of the *wave* from the motion of the *individual particles* which at any moment constitute the wave. For while the wave moves forward through great distances, the individual particles of water concerned in its propagation perform a comparatively short excursion to and fro. A sea-fowl, for example, as the waves pass it, is not carried forward, but moves up and down.*

343. Here, as in other cases, the distance through which the individual water particles oscillate, or through which the fowl moves vertically up and down, is called the *amplitude* of the oscillation.

344. When light from two different sources passes through the same ether, the waves from the one source must be more or less affected by the waves from the other. This action is most easily illustrated by reference to water-waves.

345. Let two stones be cast at the same moment into still water. Round each of them will spread a series of circular waves. Let us fix our attention on a point A in the water, equally distant from the two centres of disturbance. The two first crests of both systems of waves reach this point at the same moment, and it is lifted by their joint action to twice the height that it would attain through the action of either wave taken singly.

346. The first depression, or *sinus* as it is called, of the one system of waves also reaches the point A at the same moment as the first sinus of the other, and through their joint action the point is depressed to twice the depth that it would attain by the action of either sinus taken singly.

347. What is true of the first crest and the first depression is also true of all the succeeding ones. At the point A the successive crests will coincide, and the successive depressions will coincide, the agitation of the point being twice what it would be if acted upon by one only of the systems of waves.

* *Strictly speaking* the water particles describe *closed curves*, and not straight *vertical lines.*

348. The *length of a wave* is the distance from any crest, or any sinus, to the crest or sinus next preceding or succeeding. In the case of the two stones dropped at the same moment into still water, it is manifest that the coincidence of crest with crest and of sinus with sinus would also take place if the distance from the one stone to the point A exceeded the distance of the other stone from the same point by *a whole wave-length.* The only difference would be, that the second wave of the nearest stone would then coincide with the first wave of the most distant one. The one system of waves would here be retarded a whole wave-length behind the other system.

349. A little reflection will also make it clear that coincidence of crest with crest and of sinus with sinus will also occur at the point A when the retardation of the one system behind the other amounts to any number of *whole wave-lengths.*

350. But if we suppose the point A to be *half a wave-length* more distant from the one stone than from the other, then as the waves pass the point A the crests of one of the systems will always coincide with the sinuses of the other. When a wave of the one system tends to elevate the point A, a wave from the other system will, at the same moment, tend to depress it. As a consequence the point will neither rise nor sink, as it would do if acted upon by either system of waves taken singly. The same neutralization of motion occurs where the difference of path between the two stones and the point A amounts to any *odd* number of half wave-lengths.

351. Here, then, by adding motion to motion, we abolish motion and produce rest. In precisely the same way we can, by adding sound to sound, produce silence, one system of sound-waves being caused to neutralize another. So also by adding heat to heat we can produce cold, while by adding light to light we can produce darkness. It is this perfect identity of the deportment of light and radiant heat with the phenomena of wave-motion that constitutes the strength of the Theory of Undulation.

352. This action of one system of waves upon another, whereby the oscillatory motion is either augmented or diminished, is called *Interference.* In relation to optical phenomena it is called the Interference of Light. We shall henceforth have frequent occasion to apply this principle.

Diffraction, or the Inflexion of Light.

353. Newton, who was familiar with the idea of an ether, and indeed introduced it in some of his speculations, objected that if light were propagated by waves, shadows could not exist; for that the waves would bend round opaque bodies, and abolish the shadows behind them. According to the wave theory this bending round of the waves actually occurs, but the different portions of the inflected waves *destroy each* other by their interference.

E

354. This bending of the waves of light round the edges of opaque bodies, receives the name of *Diffraction* or *Inflexion* (German, Beugung). We have now to consider some of the effects of diffraction.

355. And for this purpose it is necessary that our source of light should be a physical point or a fine line: for when an extensive luminous surface is employed, the effects of its different points in diffraction phenomena neutralize each other.

356. A *point* of light may be obtained by converging, by a lens of short focus, the parallel rays of the sun, admitted through a small aperture into a dark room. The small image of the sun formed at the focus is here our luminous point. The image of the sun formed on the surface of a silvered bead, or indeed upon the convex surface of a glass lens, or of a watch-glass blackened within, also answers the purpose.

357. A *line* of light is obtained by admitting the sunlight through a slit, and sending the slice of light through a cylindrical lens. The rectangular beam is contracted to a physical line at the focus of the lens. A glass tube blackened within and placed in the light, reflects from its surface a luminous line which also answers the purpose. For many experiments, indeed, the circular aperture, or the slit itself, suffices without any condensation by a lens.

358. In the experiment now to be described, a slit of variable width is placed in front of the electric lamp, and this slit is looked at from a distance through another slit, also of variable aperture. The light of the lamp is rendered monochromatic by placing a pure red glass in front of the slit.

359. With the eye placed in the straight line drawn through both slits from the incandescent carbon points of the electric lamp an extraordinary appearance is observed. Firstly, the slit in front of the lamp is seen as a vivid rectangle of light; but right and left of it is a long series of rectangles, decreasing in vividness, and separated from each other by intervals of absolute darkness.

360. The breadth of the bands varies with the width of the slit placed in front of the eye. If the slit be widened the images become narrower, and crowd more closely together; if the slit be narrowed, the images widen and retreat from each other.

361. It may be proved that the width of the bands is inversely proportional to the width of the slit held in front of the eye.

362. Leaving everything else unchanged, let a blue glass or a solution of ammonia sulphate of copper, which gives a very pure blue, be placed in the path of the light. A series of blue bands is thus obtained, exactly like the former in all respects save one ; the blue rectangles are *narrower*, and they are *closer together* than the red ones.

363. If we employ colours of intermediate refrangibilities between *red and blue*, which we may do by causing the different colours of a

spectrum to shine through the slit, we should obtain bands of colour intermediate in width and occupying intermediate positions between those of the red and blue. Hence when *white light* passes through the slit the various colours are not superposed, and instead of a series of monochromatic bands, separated from each other by intervals of darkness, we have a series of coloured spectra placed side by side, the most refrangible colour of each spectrum being nearest to the slit.

364. When the slit in front of the camera is illuminated by a candle flame, instead of the more intense electric light, substantially the same effects, though less brilliant, are observed.

365. What is the meaning of this experiment, and how are the lateral images of the slit produced? Of these and certain accompanying results the emission theory is incompetent to offer any explanation. Let us see how they are accounted for by the theory of undulation.

366. For the sake of simplicity, we will consider the case of monochromatic light. Conceive a wave of ether advancing from the first slit towards the second, and finally filling the second slit. When the wave passes through the latter it not only pursues its direct course to the retina, but diverges right and left, tending to throw into motion the entire mass of the ether behind the slit. In fact, *every point of the wave which fills the slit is itself a centre of new wave-systems, which are transmitted in all directions through the ether behind the slit.* We have now to examine how these secondary waves act upon each other.

367. First, let us regard the central rectangle of the series. It is manifest that the different parts of every transverse section of the wave, which in this case fills our slit, reach the retina at the same moment. They are in complete accordance, for no one portion is retarded in reference to any other portion. The rays thus coming direct from the source through the slit to the retina produce the central band of the series.

368. But now let us consider those waves which diverge *obliquely* from the slit. In this case, the waves from the two edges of the slit have, in order to reach the retina, to pass over *unequal distances.* Let us suppose the difference in path of the two marginal rays to be a whole wave-length of the red light; how must this difference affect the final illumination of the retina?

369. Fix your attention upon the particular ray or line of light that passes exactly through the *centre* of the slit to the retina. The difference of path between this central ray and the two marginal rays is, in the case here supposed, *half a wave-length.* The least reflection will make it clear that every ray on the one side of the central line finds a ray upon the other side, from which its path differs by half an undulation, with which, therefore, it is in complete discordance. The consequence is that the light on the one side of the central line will completely abolish the light on the other side of that line, absolute darkness being the result of their mutual extinction. The first *dark*

interval of our series of bands is thus accounted for. It is produced by an obliquity which causes the paths of the marginal rays to be *a whole wave-length* different from each other.

370. When the difference between the paths of the marginal rays is *half a wave-length*, a *partial* destruction of the light is effected. The luminous intensity corresponding to this obliquity is a little less than one-half—accurately 0·4—of that of the undiffracted light.

371. If the paths of the marginal rays be three semi-undulations different from each other, and if the whole beam be divided into three equal parts, two of these parts will completely neutralize each other, the third only being effective. Corresponding, therefore, to an obliquity which produces a difference of three semi-undulations in the marginal rays, we have a luminous band, but one of considerably less intensity than the undiffracted central band.

372. With a marginal difference of path of four semi-undulations we have a second extinction of the entire beam, a space of absolute darkness corresponding to this obliquity. In this way we might proceed further, the general result being that, whenever the obliquity is such as to produce a marginal difference of path of an *even* number of semi-undulations, we have complete extinction; while, when the marginal difference is an *odd* number of semi-undulations, we have only partial extinction, a portion of the beam remaining as a luminous band.

373. A moment's reflection will make it plain that the shorter the wave, the less will be the obliquity required to produce the necessary retardation. The maxima and minima of blue light must therefore fall nearer to the centre than the maxima and minima of red light. The maxima and minima of the other colours fall between these extremes. In this simple way the undulatory theory completely accounts for the extraordinary appearance referred to in Note 359. When a slit and telescope are used, instead of the slit and naked eye, the effects are magnified and rendered more brilliant.

Measurement of the Waves of Light.

374. We are now in a condition to solve the important problem of measuring *the length* of a wave of light.

375. The first of our dark bands corresponds, as already explained, to a difference of marginal path of one undulation; our second dark band to a difference of path of two undulations; our third dark band to a difference of three undulations, and so forth. With a slit 1·35* millimeter wide Schwerd found the angular distance of the first dark band from the centre of the field to be 1′ 38″. The angular distances of the other dark bands are twice, three times, four times, &c., this quantity, that is to say they are *in arithmetical progression*.

* The millimeter is about 1/25th of an inch.

376. Draw a diagram of the slit E C with the beam passing through it at the obliquity corresponding to the first dark band. Let fall a perpendicular from one edge, E, of the slit on the marginal ray of the other edge at d. The distance, c d, between the foot of this perpendicular and the other edge is the length of the wave of light. From the centre E, with the width E C as radius, suppose a semicircle to be described; its radius being 1·35, the length of this semicircle is readily found to be 4·248 millimeters. Now, the length of this semicircle is to the length c d of the wave as 180° to 1'38", or as 648,000" to 98". Thus we have the proportion—

648,000 : 98 :: 4·248 to the wave-length c d.*

Making the calculation we find the wave-length for this particular kind of light (red), to be 0·000643 of a millimeter, or 0·000026 of an inch.

377. Instead of receiving them directly upon the retina, the coloured fringes may be received upon a screen. In this case it is desirable to employ a lens of considerable convergent power to bring the beam from the first slit to a focus, and to place the second slit or other diffracting edge or edges between the focus and the screen. The light in this case virtually emanates from the focus.

378. If the edge of a knife be placed in the beam parallel to the slit, the shadow of the edge upon the screen will be bounded by a series of parallel coloured fringes. If the light be monochromatic the bands will be simply bright and dark. The back of the knife produces the same effect as its edge. A wooden or an ivory paper-knife produces precisely the same effect as a steel knife. The fringes are absolutely independent of the character of the substance round the edge of which the light is diffracted.

379. A thick wire placed in the beam has coloured fringes on each side of its shadow. If the wire be *fine*, or if a human hair be employed, the geometric shadow itself will be found occupied by parallel stripes. The former are called the *exterior fringes*, the latter the *interior fringes*. In the hands of Young and Fresnel all these phenomena received their explanation as effects of interference.

380. A *slit* consists of two edges facing each other. When a slit is placed in the beam between the focus and the screen, the space between the edges is occupied by stripes of colour.

381. Looking at a distant point of light through a small circular aperture the point is seen encircled by a series of coloured bands. If monochromatic light be used these bands are simply bright and dark, but with white light the circles display iris-colours.

382. These results are capable of endless variation by varying the size, shape, and number of the apertures through which the

---

* C d is so minute that it practically coincides with the circle drawn round it.

point of light is observed. The street lamps at night, looked at through the meshes of a handkerchief, show diffraction phenomena. The diffraction effects obtained by Schwerd in looking through a bird's feathers are very gorgeous. The iridescence of Alpine clouds is also an effect of diffraction.*

383. Following out the indications of theory Poisson was led to the paradoxical result that in the case of *an opaque circular disk* the illumination of the centre of the shadow, caused by diffraction at the edge of the disk, is precisely the same as if the disk were altogether absent. This startling consequence of theory was afterwards verified experimentally by Arago.

Colours of Thin Plates.

384. When a beam of monochromatic light—say of pure red, which is most easily obtained by absorption—falls upon a thin, transparent film, a portion of the light is reflected at the first surface of the film; a portion enters the film, and is in part reflected at the second surface.

385. This second portion having crossed the film to and fro is *retarded* with reference to the light first reflected. The case resembles that of our two stones dropped into still water at unequal distances from the point A (Note 345).

386. If the thickness of the film be such as to retard the beam reflected from the second surface a whole wave-length, or any number of whole wave-lengths—or, in other words, any *even* number of half wave-lengths—the two reflected beams, travelling through the same ether, will be in *complete accordance*; they will therefore support each other, and make the film appear brighter than either of them would do taken singly.

387. But if the thickness of the film be such as to retard the beam reflected from the second surface half a wave-length, or any *odd* number of half wave-lengths, the two reflected beams will be in *complete discordance*; and a destruction of light will follow. By the addition of light which has undergone more than one reflexion at the second surface to the light which has undergone only one

---

* This may be imitated by the spores of Lycopodium. The diffraction phenomena of 'actinic clouds' are exceedingly splendid. One of the most interesting cases of diffraction by small particles that ever came before me was that of an artist whose vision was disturbed by vividly-coloured circles. When he came to me he was in great dread of losing his sight; assigning as a cause of his increased fear that the circles were becoming larger and the colours more vivid. I ascribed the colours to minute particles in the humours of the eye, and encouraged him by the assurance that the increase of size and vividness indicated that the diffracting particles were becoming *smaller*, and that they might finally be altogether absorbed. The prediction was verified. It is needless to say one word on the necessity of optical knowledge in the case of the practical oculist.

reflexion, the beam reflected from the first surface may be *totally* destroyed. Where this total destruction of light occurs the film appears black.

388. If the film be of variable thickness, its various parts will appear bright or dark according as the thickness favours the accordance or discordance of the reflected rays.

389. Because of the different lengths of the waves of light, the different colours of the spectrum require different thicknesses to produce accordance and discordance; the longer the waves, the greater must be the thickness of the film. Hence those thicknesses which effect the extinction of one colour will not effect the extinction of another. When, therefore, a film of variable thickness is illuminated by *white* light, it displays a variety of colours.

390. These colours are called the colours of *thin plates.*

391. The colours of the soap-bubble; of oil or tar upon water; of tempered steel; the brilliant colours of lead skimmings; Nobili's metallo-chrome; the flashing colours of certain insects' wings, are all colours of thin plates. The colours are produced by transparent films of all kinds. In the bodies of crystals we often see iridescent colours due to vacuous films produced by internal fracture. In cutting the dark ice under the moraines of glaciers internal fracture often occurs, and the colours of thin plates flash forth from the body of the ice with extraordinary brilliancy.

392. Newton placed a lens of small curvature in optical contact with a plane surface of glass. Between the lens and the surface he had a film of air, which gradually augmented in thickness from the point of contact outwards. He thus obtained in monochromatic light a series of bright and dark *rings*, corresponding to the different thicknesses of the film of air, which produced alternate accordance and discordance.

393. The rings produced by violet he found to be smaller than those produced by red, while the rings produced by the other colours fell between these extremes. Hence when white light is employed, ' Newton's Rings' appear as a succession of circular bands of colour. A far greater number of the rings is visible in monochromatic than in white light, because the differently coloured rings, after a certain thickness of film has been attained, become superposed and re-blended to form white light.

394. Newton, considering the means at his disposal, measured the diameters of his rings with marvellous accuracy; he also determined from its focal length and its refractive index the diameter of the sphere of which his lens formed a part. He found the squares of the diameters of his rings to be in *arithmetical progression*, and consequently that the *thicknesses* of the film of air corresponding to the diameters of the rings were also in arithmetical progression.

395. He determined the *absolute thicknesses* of the plates of air at which the rings were formed. Employing the most luminous rays of

the spectrum, that is the rays at the common boundary of the yellow and orange, he found the thickness corresponding to the first bright ring to be $\frac{1}{178000}$th of an inch.

396. The entire series of bright rings were formed at the following successive thicknesses :—

$$\frac{1}{178000}, \frac{3}{178000}, \frac{5}{178000}, \frac{7}{178000}, \&c.,$$

and the series of dark rings, separating the bright ones, at the thicknesses

$$\frac{2}{178000}, \frac{4}{178000}, \frac{6}{178000}, \frac{8}{178000}, \&c.$$

397. To account for the rings, Newton assumed that the light particles were endowed with *fits of easy transmission and of easy reflexion.* He probably figured those particles as endowed at the same time with a motion of translation through space, and a motion of rotation round their own axes. If we suppose such particles to resemble little magnets which present alternately attractive and repulsive poles to the surface which they approach, we have a conception in conformity with the notion of Newton.

398. According to this conception ordinary reflexion and refraction would depend upon the presentation of the repulsive or the attractive poles of the particles to the reflecting or refracting surface.

399. Figure then the rotating light particles entering the film of air between Newton's lens and plate. If the distance between both be such as to enable the light particle to perform a *complete rotation*, it will present *at the second surface* of the film of air the same pole that it presented at the first. It will therefore be *transmitted*, and will not return to the eye.

400. This effect would also take place if the distance between the plate and lens were such as to enable the light particle to perform two, three, four, &c., complete rotations. *The dark rings of Newton were thus accounted for.* They occurred at places where the light particles, instead of being sent back to the eye from the second surface of the film, were transmitted through that surface.

401. But if the thickness of the film be such as to allow the light particle which has entered the first surface to perform only *half a rotation* before it arrives at the second surface ; then a repulsive pole will be presented to the latter, and the particle will be driven back to the eye. The same will occur if the distance be such as to enable the light particle to perform three, or five, or seven, &c., semi-rotations. *The bright rings of Newton were thus accounted for;* they occurred at places where the light particles on reaching the second surface of the film were reflected back to the eye.

402. The theory of emission is here at direct issue with the theory of undulation. Newton assumes that the action which produces the alternate bright and dark rings takes place at a *single* surface; *i. e.* the second surface of the film. The undulatory theory affirms that the rings are caused by the interference of rays reflected from *both*

surfaces. This has been proved to be the case. By employing polarised light (to be subsequently described and explained) we can destroy the reflexion at the first surface of the film, and when this is done the rings vanish altogether.

403. The beauty and subtlety of Newton's conception are, however, manifest; and the theory was apparently supported by the fact that rings of feeble intensity are actually formed by *transmitted light*, and that the bright rings by transmitted light correspond to thicknesses which produce dark rings in reflected light.

404. The transmitted rings are referred by the undulatory theory to the interference of rays which have passed directly through the film, with others which have undergone *two reflexions* within the film. They are thus completely accounted for.

Note.—The thickness $\frac{1}{178000}$ of an inch referred to in Note 396, as that corresponding to the first bright ring, is *one-fourth* of the length of an undulation of the light employed by Newton. Hence, in passing to and fro through the film, the rays reflected at the second surface are *half* an undulation behind those reflected at the first surface. At this thickness, therefore, the ring ought, according to the principles of interference, to be *dark* instead of *bright*. The same remarks apply to the thicknesses $\frac{3}{178000}$, $\frac{5}{178000}$, &c.; the former corresponds to a retardation of three, and the latter to a retardation of five semi-undulations. With regard to the *dark* rings, the first of them occurs at a thickness the double of which is the length of a whole undulation; the second of them occurs at a thickness which, when doubled, is equal to two wave-lengths; the third at a thickness double of which is three wave-lengths. Hence, if we take *the thickness of the film alone* into account, the bright rings ought to be dark, and the dark rings bright.

But something besides thickness is to be considered here. In the case of the first surface of the film the wave passes from the dense ether of the glass into the rare ether of the air. In the case of the second surface of the film the wave passes from the rare ether of the air into the dense ether of the glass. This difference at the two reflecting surfaces of the film can be proved to be equivalent *to the addition of half a wave-length* to the thickness of the film. To the absolute thickness, therefore, as measured by Newton, half a wave-length is in each case to be added; when this is done the rings follow each other in exact accordance with the law of interference enunciated in Notes 348 to 350.

Double Refraction.

405. In air, water, and well-annealed glass, the luminiferous ether has the same elasticity in all directions. There is nothing in the molecular grouping of these substances to interfere with the perfect homogeneity of the ether.

406. But when water crystallizes to ice, the case is different; here

the molecules are constrained by their proper forces to arrange themselves in a certain determinate manner. They are, for example, closer together in some directions than in others. This arrangement of the molecules carries along with it an arrangement of the surrounding ether, which causes it to possess *different degrees of elasticity in different directions.*

407. In a plate of ice, for example, the elasticity of the ether in a direction perpendicular to the surface of freezing is different from its elasticity in a direction parallel to the same surface.

408. This difference is displayed in a peculiarly striking manner by Iceland spar, which is crystallized carbonate of lime; and in consequence of the existence of these two different elasticities, a wave of light passing through the spar *is divided into two*; the one rapid, corresponding to the greater elasticity, and the other slow, corresponding to the lesser elasticity.

409. Where the velocity is greatest, the refraction is least; and where the velocity is least the refraction is greatest. Hence in Iceland spar, as we have two waves moving with different velocities, we have *double refraction.*

410. This is also true of the greater number of crystalline bodies. If the grouping of the molecules be not in all directions alike, the ether will not be in all directions equally elastic, and double refraction will infallibly result.

411. In rock salt, alum, and other crystals this homogeneous grouping of the molecules actually occurs, and such crystals behave like glass, water, or air.

412. In certain doubly refracting crystals the molecules are arranged in the same manner on all sides of a certain direction. For example, in the case of ice the molecular arrangement is the same all round the perpendiculars to the surface of freezing.

413. In like manner, in Iceland spar the molecules are arranged symmetrically round the crystallographic axis, that is, round the shortest diagonal of the rhomb into which the crystal may be cloven.*

414. When a beam of light passes through ice perpendicular to the surface of freezing, or through Iceland spar parallel to the crystallographic axis, *there is no double refraction.* These cases are representative; that is to say, there is no double refraction in the direction round which the molecular arrangement is in all directions the same.

---

* The arrangement of the molecules is such, that Iceland spar may be cloven with great and equal facility in three different directions. The *planes of cleavage* are here oblique to each other. Rock salt also cleaves readily and equally in three directions, the planes of cleavage being at right angles to each other. Hence, while rock salt cleaves into *cubes*, Iceland spar cleaves into *rhombs.* Many crystals cleave with different facilities in different directions. Selenite and *crystallized sugar* (sugar-candy) are examples.

415. This direction of no double refraction is called *the optic axis* of the crystal.

NOTE.—The vibrations of the ether being *transverse* to the direction of the ray, the elasticity which determines the rapidity of transmission is that *at right angles* to the ray's direction. In Iceland spar the velocity is slowest in the direction of the axis; hence the elasticity at right angles to the axis is a *minimum*. The ray, on the other hand, whose vibrations are executed along the axis is the most rapid; hence the elasticity of the ether along the axis is a *maximum*. In perfectly homogeneous bodies the surface of elasticity would be spherical; it would be measured by the same length of radius in all directions. In the case of Iceland spar the surface of elasticity is an ellipsoid whose longer axis coincides with the axis of the crystal.

Phenomena presented by Iceland Spar.

416. The two beams into which the incident beam is divided by the spar do not behave alike. One of them obeys the ordinary law of refraction; its index of refraction is perfectly constant and independent of its direction through the crystal. The angles of incidence and refraction are in the same plane, as in the case of ordinary refraction. The ray which behaves thus is called *the ordinary ray*. In its case the sine of the angle of incidence is to the sine of the angle of refraction, or the velocity of light in air is to its velocity in the crystal, in the constant ratio of 1·654 to 1. The number 1·654 is *the ordinary index* of Iceland spar.

417. But the other beam acts differently. Its index of refraction is not constant, nor is the angle of refraction as a general rule in the same plane as the angle of incidence. The ray which behaves thus is called *the extraordinary ray*. If a prism be formed of the spar with its refracting angle parallel to the optic axis, when the incident beam traverses the prism *at right angles to the optic axis*, the separation of its two parts is a *maximum*. Here the full difference of elasticity between the axial direction and that perpendicular to it comes into play, and the extraordinary ray suffers its *minimum* retardation, and therefore its minimum refraction. Its refractive index is then 1·483.

418. The index of refraction of the extraordinary ray varies with its direction through the crystal from 1·483 to 1·654. The *minimum value* of the ratio of the two sines, or of the two velocities, *viz.* 1·483, is called *the extraordinary index*.

419. When a small aperture through which light passes is regarded through a rhomb of Iceland spar two apertures are seen. If the rhomb be placed over a black dot on a sheet of white paper, two dots will be seen; and if the spar be turned, one of the images of the aperture or of the dot will rotate round the other.

420. The rotating image is that formed by the extraordinary ray.

421. One of the two images of the dot is also *nearer* than the other. The ordinary ray behaves as if it came from a more highly refractive medium, and the greater the refraction the nearer must the image appear. The apparent shallowness of water is referred to in Notes 131 and 132. With bisulphide of carbon the shallowness would be more pronounced, because the refraction is greater. In Iceland spar the ordinary index bears nearly the same relation to the extraordinary as the index of bisulphide of carbon to that of water; hence the *ordinary image* must appear nearer than the extraordinary one.

422. Brewster showed that a great number of crystals possessed *two* optic axes, or two directions on which a beam passes through the crystal without division. Crystallized sugar, mica, heavy spar, sulphate of lime and topaz are examples.

423. Thus crystals divide themselves into—

I. *Single refracting crystals*, such as rock salt, alum, and fluor spar; and

II. *Double refracting crystals*, of which we have two kinds, *viz.*

a. Uniaxal crystals, or those with a single optic axis, such as Iceland spar, rock crystal, and tourmaline; and

b. Biaxal crystals, or those which possess two optic axes, such as arragonite, felspar, and those mentioned in 422.

424. When on a plate of Iceland spar cut perpendicular to the axis, a beam of light falls obliquely, the ordinary ray being the more refracted is nearer to the axis than the extraordinary. The extraordinary ray is as it were *repelled* by the axis. But Biot showed that there are many crystals in which the reverse occurs, in which, that is to say, the extraordinary ray is nearer to the axis than the ordinary, being as it were *attracted*. The former class he called repulsive or *negative* crystals; Iceland spar, ruby, sapphire, emerald, beryl, and tourmaline being examples. The latter class he called attractive or *positive* crystals, rock crystal, ice, zircon being examples.

The Polarization of Light.

425. The double refraction of Iceland spar was discovered by Erasmus Bartholinus, and was first described by him in a work published in Copenhagen in 1669. The celebrated Huygens sought to account for the phenomenon on the principles of a wave theory, and he succeeded in doing so.

426. In his experiments on this subject, Huygens found that when a common luminous beam passes through Iceland spar in any direction save one (that of the optic axis), it is always divided into two beams of *equal intensity*; but that when *either* of these two half-beams is sent through a second piece of spar, it is usually divided into two of *unequal intensity*; and that there are two positions of the spar in which one of *the beams* vanishes altogether.

427. On turning the spar round this position of absolute disappearance, the missing beam appeared; its companion at the same time becoming dimmer; both of them then passed through a phase of equal intensity, and when the rotation was continued, the beam which was first transmitted disappeared.

428. Reflecting on this experiment Newton came to the conclusion, that the divided beam had acquired *sides* by its passage through the Iceland spar, and that its interception and transmission depended on the way on which those sides presented themselves to the molecules of the second crystal. He compared this *two-sidedness* of a beam of light to the *two-endedness* of a magnet known as its polarity; and a luminous beam exhibiting this two-sidedness was afterwards said to be *polarized*.

429. In 1808, Malus, while looking through a birefracting prism at one of the windows of the Luxembourg Palace, from which the solar light was reflected, found that in a certain position of the spar, the ordinary image of the window almost wholly disappeared; while in a position perpendicular to this, the extraordinary image disappeared. He discerned the analogy between this action and that discovered by Huygens in Iceland spar, and came to the conclusion that the effect was due to some new property impressed upon the light by its reflexion from the glass.

430. What is this property? It may be most simply studied and understood by means of the crystal called tourmaline. This crystal is birefractive; it divides a beam of light incident upon it into two, but its molecular grouping, and the consequent disposition of the ether within it, are such that one of these beams is rapidly quenched, while the other is transmitted with comparative freedom.

431. It is to be borne in mind that the motions of the individual ether particles are transverse to the direction in which the light is propagated (read Note 219). *In a beam of ordinary light the vibrations occur in all directions round the line of propagation.*

432. The change suffered by light in passing through a plate of tourmaline, of sufficient thickness, and cut parallel to the axis is this:—All vibrations save those executed *parallel to the axis* are quenched within the crystal. Hence the beam emergent from the plate of tourmaline has all its vibrations reduced to a single plane. In this condition it is a beam *of plane polarized light.*

433. Imagine a cylindrical beam of light with all its ether particles vibrating in the same direction—say *horizontally*—looked down upon vertically, the ether particles, if large enough, would be seen performing their excursions to and fro across the direction of the beam. Looked at crosswise horizontally, the particles would be seen advancing and retreating, but their paths would be invisible, every ether particle covering its own path. In the one case we should see *the lines* of excursion; in the other case, *the ends* of the lines only. In this, according to the undulatory theory, consists the *two-sidedness* discovered by Huygens, and commented on by Newton.

Polarization of Light by Reflexion.

434. The quality of two-sidedness is also impressed upon light by reflexion. This is the great discovery of Malus. A beam reflected from glass is in part polarized *at all oblique incidences,* a portion of its vibrations being reduced to a common plane. At one particular incidence the beam is *perfectly polarized, all* its vibrations being reduced to the same plane. The angle of incidence which corresponds to this perfect polarization is called the *polarizing angle.*

435. The polarizing angle is connected with the index of refraction of the medium by a very beautiful law discovered by Sir David Brewster.* When a luminous beam is incident upon a transparent substance, it is in part reflected and in part refracted. At one particular incidence the reflected and refracted portions of the beam are *at right angles to each other*. The angle of incidence is *then* the polarizing angle. This is the geometrical expression of the law of Brewster.

436. The polarizing angle augments with the refractive index of the medium. For water it is 53°, for glass 58°, and for diamond 68°.

437. Thus a beam of ordinary light, whose vibrations are executed in all directions, impinging upon a plate of glass at the polarizing angle, has, after reflexion, all its vibrations reduced to a common plane. The direction of the vibrations of the polarized beam *is parallel to the polarizing surface.*

438. Let a beam thus polarized by reflexion at the surface of one plate of glass impinge upon a second plate *at the polarizing angle.* In one position of this plate the beam suffers its maximum reflexion. In a certain other position the beam is *wholly transmitted*, there is no reflexion. In this experiment the angle of incidence remains unchanged, nothing being altered save *the side* of the ray which strikes the reflecting surface.

439. The reflexion of the polarized beam is a maximum when the lines along which the ether particles vibrate are *parallel* to the reflecting surface. It is wholly transmitted when the lines of vibration strike the reflecting surface at the polarizing angle. The reflexion is then zero. By taking advantage of this fact, the reflexion from the first surface of a thin film has been abolished, Newton's rings being thereby rendered incapable of formation, as stated in Note 402.

440. A beam which meets the first surface of a plate of glass with parallel sides at the polarizing angle meets the second surface also at its polarizing angle, and is in part reflected there perfectly polarized. Hence, by augmenting the number of plates, the

* The index of refraction of the medium is the tangent of the polarizing angle.

repeated reflexions at their limiting surfaces furnish a polarized beam of greater intensity than that obtained by reflexion at a single surface.

Polarization of Light by Refraction.

441. We have hitherto directed our attention to the *reflected* portion of the beam; but the *refracted* portion, which enters the glass, is also partially polarized. The quantities of polarized light in the reflected and refracted beams *are always equal to each other.*

442. The plane of vibration in the refracted beam *is at right angles* to that in the reflected beam.

443. When several plates of glass are placed parallel to each other, and a beam is permitted to fall upon them at the polarizing angle, at every passage from plate to plate a portion of light is reflected polarized, an equal portion of polarized light entering the glass at the same time. By duly augmenting the number of plates, the polarization by the successive refractions may be rendered sensibly *perfect.* When this occurs, if any further plates be added to the bundle, reflexion *entirely ceases* at their limiting surfaces, the beam afterwards being wholly transmitted.

Polarization of Light by Double Refraction.

444. In the case last considered the light was polarized by ordinary refraction. The polarization of light by double refraction has been already touched upon in Notes 432 and 433. We shall now extend our examination of the crystal of tourmaline there referred to, and turn it to account in the examination of other crystals.

445. If a beam of light which has passed through one plate of tourmaline impinge upon a second plate, it will pass through both, if the axes of the two plates be *parallel.* But if they are *perpendicular* to each other, then the light transmitted by the one is quenched by the other, darkness marking the space where the two plates are superposed.

446. If the two axes be *oblique* to each other, a portion of the light will pass through both plates. For, in a manner similar to the resolution of forces in ordinary mechanics, an oblique vibration may be resolved into two, one parallel to the axis of the tourmaline, the other perpendicular to the axis. The latter component is *quenched*, but the former is *transmitted.*

447. Hence if the axes of two plates of tourmaline be perpendicular to each other, a third plate of tourmaline introduced *obliquely* between them, or a plate of any other crystal which acts in a manner similar to the tourmaline, will transmit a portion of the light emergent from the first crystal. The plane of vibration of this light being oblique to the axis of the second crystal, a portion of the light will also pass through the latter. By the intro-

duction, therefore, of a third crystal, with its axis oblique, we abolish in part the darkness of the space where the two rectangular plates are superposed.

Examination of Light transmitted through Iceland Spar.

448. We have now to examine, by means of a plate of tourmaline, the two parts into which a luminous beam is divided in its passage through Iceland spar.

449. Confining our attention to one of the two beams, it is immediately found that in a certain position of the plate the light is freely transmitted, while in the perpendicular position it is completely stopped. This proves the beam emergent from the spar to be *polarized*.

450. From the position of the tourmaline we can immediately infer the direction of vibration in the polarized beam. If transmission occur when the axis of the plate of tourmaline is vertical, the vibrations are vertical; if transmission occur when the tourmaline is horizontal, the vibrations are horizontal. The same mode of investigation teaches us that the second beam emergent from the spar is also polarized.

451. The vibrations of the ether particles in the two beams are executed in planes which are *at right angles to each other*. If the vibrations in the one beam be vertical, in the other they are horizontal. A plate of tourmaline with its axis vertical transmits the former and quenches the latter; while the same plate held horizontally, quenches the former and transmits the latter.

452. A tourmaline plate placed with its axis vertical, in front of the electric lamp, has its image cast by a lens upon a screen. A piece of Iceland spar, with one of its planes of vibration horizontal and the other vertical, placed in front of the lens divides the beam into two, and yields *two images* of the tourmaline. One of these images is *bright*, the other is *dark*. The reason is that in the light emergent from the tourmaline the vibrations are vertical, and they can only be transmitted through the spar in company with *its* vertically vibrating beam. In the horizontally vibrating beam the tourmaline must appear black.

453. It is also black if the light emergent from it, and surrounding it, meet, at the polarizing angle, a plate of glass whose plane of reflexion is *vertical*; while it is bright when the light is reflected *horizontally*. These effects are consequences of the law of polarization by reflexion.

454. Not only do crystallized bodies possess this power of double refraction and polarization; but all bodies whose atomic grouping is such as to cause the ether within them to possess different elasticities in different directions do the same.

455. Thus organic structures are usually double refracting. A

double refracting structure may also be conferred on ordinary glass by either strain or pressure. Strains and pressures due to unequal heating also produce double refraction. Unannealed glass behaves like a crystal. A plate of common window-glass, which under ordinary circumstances shows no trace of double refraction, if heated at a single point, is rendered doubly refractive by the strains and pressures propagated round the heated point. The introduction of any of these bodies between the *crossed plates* of tourmaline partly abolishes the darkness caused by the superposition of the plates.

456. Two plates of tourmaline, between which bodies may be introduced and examined by polarized light, constitute a simple form of the *polariscope*. The plate at which the light first enters is called the polarizer, while the second plate is called the analyzer.

457. But the tourmalines are small, usually coloured, and under no circumstances competent to furnish an intense beam of polarized light. If one of the parts into which a prism of Iceland spar divides a beam of light could be abolished, the remaining beam would be polarized, and, because of the transparency of the spar, it would be far more intense than any beam obtainable from tourmaline.

458. This has been accomplished with great skill by Nicol. He cut a long parallelopiped of spar into two by a very oblique section; polished the two surfaces, and united them by Canada balsam. The refrangibility of the balsam lies between those of the ordinary and the extraordinary rays in Iceland spar, being less than the former and greater than the latter. When, therefore, a beam of light is sent along the parallelopiped, the ordinary ray, to enter the balsam, must pass from *a denser to a rarer medium*. In consequence of the obliquity of its incidence *it is totally reflected*, and is thus got rid of. The extraordinary ray, on the contrary, in passing from the spar to the balsam passes from a rarer to a denser medium, and is therefore *transmitted*. In this way we obtain a single intense beam of polarized light. (Read Notes 123, 141, and 142.)

459. A parallelopiped prepared in the fashion here described is called a *Nicol's prism*.

460. Nicol's prisms are of immense use in experiments on polarization. With them the best polariscopes are constructed. Reflecting polariscopes are also constructed, consisting of two plates of glass, one of which polarizes the light by reflexion, the other examining the light so polarized. The beam reflected from the polarizer is in this case reflected or quenched by the analyzer according as the planes of reflexion of the two mirrors are parallel or at right angles to each other.

Colours of Double-refracting Crystals in Polarized Light.

461. A large class of these colours may be illustrated and explained by reference to the deportment of thin plates of gypsum

F

(crystallized sulphate of lime, commonly called selenite) between
the polarizer and analyzer of the polariscope.

462. The crystal cleaves with great freedom in one direction; it
cleaves with less freedom in two others; the latter two cleavages are
also unequal. In other words, gypsum possesses three planes of
cleavage, no two of which are equal in value, but one of which
particularly signalizes itself by its perfection.

463. By following these three cleavages it is easy to obtain from
the crystal diamond-shaped laminæ of any required thinness.

464. The crystal, as might be expected from the character of its
cleavages, is double-refracting. A beam of ordinary light impinging at right angles on a plate of gypsum, whose surfaces are those of
most perfect cleavage, has its vibrations reduced to two planes at
right angles to each other; that is to say, the beam whose ether,
prior to entering the gypsum, vibrates in all transverse directions,
after it has entered the gypsum, and after its emergence from it,
vibrates in two rectangular directions only.

465. The elasticity of the ether is different in these two rectangular directions; consequently the one beam passes more rapidly
through the gypsum than the other.

466. In refracting bodies generally the retardation of the light
consists in a *diminution of the wave-length* of the light. *The rate of
vibration* is unchanged during the passage of the light through the
refracting body. The case is exactly similar to that of a musical
sound transmitted from water into air. The velocity is reduced to
one-fourth by the transfer, because the wave-length is reduced to
one-fourth. But the *pitch*, depending as it does on the number of
waves which reach the ear in a second, is unaltered.

467. Because of the difference of elasticity between the two rectangular directions of vibration in gypsum, the waves of ether in the
one direction *are more shortened* than in the other.

468. In the experiments with a plate of gypsum now to be described and explained, we shall employ as polarizer a piece of Iceland spar, one of whose beams is intercepted by a diaphragm. A
Nicol's prism shall be our analyzer.

469. When the planes of vibration of the spar and of the Nicol
coincide, the light passes through both and may be received upon a
screen. When the planes of vibration are at right angles to each
other, the light emergent from the spar is intercepted by the Nicol,
and the screen is dark.

470. If a plate of selenite be placed between the polarizer and
analyzer, with either of its planes of vibration *coincident* with that of
the polarizer or analyzer, it produces no change upon the screen.
If the screen be light, it remains light; if it be dark, it remains
dark after the introduction of the gypsum, which here behaves like
a plate of ordinary glass.

471. *Let us assume the screen to be dark. Interposing a thick*

plate of gypsum with its directions of vibration *oblique* to that of the polarizer or analyzer, *white* light reaches the screen. If the plate be *thin*, the light which reaches the screen is *coloured*. If the plate be of uniform thickness, the colour is uniform. If of different thicknesses, or if in cleaving thin scales cling to the surface of the film, some portions of the plate will be differently coloured from the rest.

472. When thick plates are employed, the different colours, as in the case of thin plates, are superposed, and re-blended to white light.

473. The quantity of light which reaches the eye is a maximum when the planes of vibration of the gypsum enclose an angle of 45° with those of the polarizer and analyzer.

474. If the plate of selenite be a thin wedge, and if the light be monochromatic, say red, alternately bright (red) and dark bands are thrown upon the screen.

475. If, instead of red light, *blue* be employed, the blue bands are found to occur at smaller thicknesses than those which produced the red : other colours occur at intermediate thicknesses. Hence when *white* light is employed, instead of bands of brightness separated from each other by bands of darkness, we have a series of iris-coloured bands.

476. If, instead of a wedge gradually augmenting in thickness from the edge towards the back, we employ a disk gradually augmenting in thickness from the centre outwards; instead of a series of parallel bands we obtain under similar circumstances, in *white* light, a series of concentric iris-coloured circles.

477. Here then we have in the first instance a beam of plane polarized light impinging on the selenite. The direction of vibration of this beam is resolved into two others at right angles to each other; namely, into the two directions in which the ether vibrates within the crystal. One of these systems of waves is *retarded* with reference to the other.

478. But as long as the rays vibrate *at right angles to each other*, they cannot interfere so as to augment or diminish the intensity. To effect such interference the rays must vibrate *in the same plane.*

479. The function of the analyzer is to reduce the two rectangular wave-systems to a single plane. Here the effect of retardation is at once felt, and the waves conspire or oppose each other according as their vibrations are *in the same phase* or in *opposite phases.*

480. When the vibration planes of the polarizer and analyzer are *parallel*, a thickness of the gypsum crystal which produces a retardation of *half an undulation* causes the light to be extinguished by the analyzer.

481. When the polarizer and analyzer are *crossed*, a retardation

F 2

of half an undulation, or of any odd number of half undulations, within the crystal does not produce extinction when these vibrations are compounded by the analyzer. A retardation of a whole undulation, or of any number of whole undulations, produces in this case extinction. This, when followed out, is a plain consequence of the composition of the vibrations.

482. Expressed generally, the phenomena exhibited by the parallel and crossed polarizer and analyzer are *complementary*. If the field be dark when they are crossed, it is bright when they are parallel. If the field be green when they are crossed, it is red when they are parallel; if yellow when they are crossed, it is blue when they are parallel. Thus a rotation of 90° always brings out the complementary colour.

483. If instead of the Nicol we employ a birefracting prism of Iceland spar, the colours of the selenite produced by the two oppositely polarized beams will be complementary. The overlapping of the two colours always produces *white*. Any other double-refracting substance, whether crystallized, organized, mechanically pressed or strained, exhibits, on examination by polarized light, phenomena similar to those of the gypsum.

484. A common beam of light is equivalent in all its effects to two beams vibrating in two rectangular planes. As two such beams cannot interfere, we cannot have the colours of the selenite in common light.

Rings surrounding the Axes of Crystals in Polarized Light.

485. A pencil of rays passing along the axis through Iceland spar suffers no division; but if inclined to the axis, however slightly, the pencil is divided into two, which vibrate in rectangular planes, and one of which is more retarded than the other.

486. If the incident light be polarized, on quitting the spar, oblique to the axis, it will be in a condition similar to the light emergent from the plates of gypsum already referred to. When two rectangular vibrations, passing through the same ether, are reduced to the same plane by the analyzer, interference occurs; the two rays either conspiring or opposing each other.

487. Whether they conspire or not depends upon the amount of relative retardation, and this again depends upon the thickness of the spar traversed by the two rays. If they conspire at a certain thickness they will also conspire at twice that thickness, thrice that thickness, &c. Those thicknesses at which the rays conspire are separated by others at which they oppose each other.

488. With a conical beam whose central ray passes *along* the axis, the effects are symmetrical all round the axis; and when the crystal, illuminated by such a ray, is examined by monochromatic polarized *light, we have* a series of bright and dark circles surrounding the axis.

489. When the light is red the circles are larger than when the light is blue; the smaller the wave-length the smaller are the circles. Hence, since the different colours are not superposed, when *white* light is employed instead of bands of alternate brightness and darkness we have a series of *iris-coloured circles.*

When the polarizer and analyzer are crossed the system of bands is intersected by *a black cross,* whose arms are parallel to the planes of vibration in the polarizer and analyzer. Those rays, whose planes of vibration within the crystal coincide with the planes of either the polarizer or analyzer, *cannot get through either,* and their complete interception forms the two arms of the cross. Those rays whose planes of vibration enclose an angle of 45° with that of the polarizer or analyzer produce the greatest effect when they conspire. At this inclination the bright ring is at its maximum brilliancy, from which, right and left, it becomes more feeble, until it finally merges into the darkness of the cross.

490. A rotation of 90° produces here, as in other cases, the complementary phenomena: the black cross becomes white, and the circles change their tints to complementary ones.

491. In crystals possessing two optic axes a series of iris-coloured bands surround both axes, each band forming a curve, which its discoverer, James Bernoulli, called a *lemniscata.*

Elliptic and Circular Polarization.

492. Two rays of light vibrating *at right angles to each other,* however the one system of vibrations may be retarded with reference to the other, cannot, as already stated, interfere so as to produce either an increase or a diminution of the light.

493. But though the *intensity* remains unchanged, the rays act upon each other. If one of them differs from the other by any exact number of semi-undulations, the two rays are compounded to a single *rectilinear* vibration. In all other cases the resultant vibration is *elliptical*; in one particular case the ellipse in which the individual particles of ether move is converted into a *circle.* This occurs when one of the systems of waves is an exact quarter of an undulation behind the other; we have then *circular polarization.*

494. This compounding of ethereal vibration is mechanically the same as the compounding of the vibrations of an ordinary pendulum; or as the compounding of the vibrations of two rectangular tuning-forks by the method of Lissajous.*

495. Elliptic polarization is the *rule* and not the *exception.* It is particularly manifested in reflexion from metals, and from transparent bodies which possess a high index of refraction. Jamin has detected it in light reflected from all bodies.

* See *Lectures on Sound,* 1st ed., p. 307.

Rotatory Polarization.

496. A polarized ray of monochromatic light, as already stated, suffers no change during its transmission through Iceland spar in the direction of the optic axis.

497. But if transmitted through *rock-crystal* (quartz) in the direction of the optic axis, its plane of vibration is turned by the crystal. Supposing the polarizer and analyzer of the polariscope to be crossed so as to produce perfect darkness before the crystal is introduced between them, on its introduction light will pass, and to quench the light the analyzer must be turned into a new position. The angle through which the analyzer is turned measures the *rotation of the plane of vibration*.

498. Some specimens of rock-crystal turn the plane of vibration to the right, and others to the left. The former are called *right-handed* and the latter *left-handed* crystals. Sir John Herschel connected this optical difference with a visible difference of crystalline form.

499. In the celebrated experiment of Faraday, with a bar of heavy glass, the plane of vibration was caused to rotate both by a magnet and an electric current; the direction of rotation bearing a constant relation to the polarity of the magnet and to the direction of the current.

500. The subject of rotatory polarization was examined with great care and completeness by Biot, and he established certain laws regarding it, two of which may be enunciated here.

1. The amount of the rotation is proportional to the thickness of the plate of rock-crystal.

2. The rotation of the plane of vibration is different for the different rays of the spectrum, increasing with the refrangibility of the light.

Thus with a plate of rock-crystal one millimeter thick, he obtained the following rotations for the mean rays of the respective colours of the spectrum.

Red, 19°.	Green, 28°.	Indigo, 36°.
Orange, 21°.	Blue, 32°.	Violet, 41°.
Yellow, 23°.		

With a plate *two* millimeters in thickness the rotation for red is 38° and for violet 82°.

501. Since, then, the rays of different colours emerge from the rock-crystal vibrating in different planes, when such light falls upon the analyzer that colour only whose plane of vibration coincides with that of the analyzer will be *transmitted*. By turning the analyzer we allow the other colours to pass in succession.

502. The phenomena of rotatory polarization are produced by the *interference* of two *circularly polarized pencils* of light, which are propagated along the axis with unequal velocities, the one revolving from left to right, and the other revolving in the opposite direction.*

CONCLUSION.

I have endeavoured in these lectures to bring before you the views at present entertained by all eminent scientific thinkers regarding the nature of light. I have endeavoured to make as clear to you as possible that bold theory according to which space is filled with an elastic substance capable of transmitting the motions of light and heat. And consider how impossible it is to escape from this or some similar theory,—to avoid ascribing to light, in space, *a material basis*. Solar light and heat require about eight minutes to travel from the sun to the earth. During this time the light and heat are detached from both. Enclose, in idea, a portion of the intervening space—say a cubic mile of it—occupied for a moment by light and heat. Ask yourselves what they are. The first inquiry towards a solution is, *What can they do?* We only know things by their *effects*. What, then, are the effects which this cubic mile of light and heat can produce? At the earth, where we can operate upon them, we find them capable of producing *motion*. We can lift weights with them; we can turn wheels with them; we can urge locomotives with them; we can fire projectiles with them. What other conclusion can you come to than that the light and heat which thus produce motion *are themselves motions*? †

Our cubic mile of space, then, is for a measurable time the vehicle of motion. But is it in the human mind to imagine motion without at the same time imagining something moved? Certainly not. The very conception of motion necessarily includes that of a moving body. What, then, is the thing moved in the case of our cubic mile of sunlight? The undulatory theory replies that it is a substance of determinate mechanical properties, a body which may or may not be a form of ordinary matter, but to which, whether it is or not, we give the name of *ether*. Let us tolerate no vagueness here; for the greatest disservice that could be done to science—the surest way to give error a long lease of life—is to enshroud scientific theories in vagueness. The motion of the ether communicated to material substances throws them into motion. It is therefore itself *a material substance*, for we have no knowledge that in nature anything but a material substance can throw other material substances into motion. Two modes of motion are possible to the ether. Either it is shot through space as a *projectile*, or it is the vehicle of *wave-motion*. The projectile theory, though enunciated by Newton, and

---

* See Lloyd, *Wave Theory*, p. 199, &c.

† Sir William Thomson has attempted to calculate 'the mechanical value of a cubic mile of sunlight.'

supported by such men as Laplace, Biot, Brewster, and Malus, has hopelessly broken down. Wave-motion, then, of one kind or another we must fall back upon. But how does the Wave Theory account for the phenomena? Throughout the greater part of these lectures we have been answering this question. The cases brought before you are *representative*. Thousands of facts might be cited in illustration of each of them, and not one of these facts is left unexplained by the undulatory theory. It accounts for all the phenomena of reflexion; for all the phenomena of refraction, single and double; for all the phenomena of dispersion; for all the phenomena of diffraction; for the colours of thick plates and thin, as well as for the colours of all natural bodies. It accounts for all the phenomena of polarization; for all those wonderful affections, those chromatic splendours exhibited by crystals in polarized light. Thousands of isolated facts might, as I have said, be ranged under each of these heads; the undulatory theory accounts for them all. It traces out illuminated paths through what would otherwise be the most hopeless jungle of phenomena in which human thought could be involved. This is why the foremost men of the age accept the ether not as a vague dream, but as a real entity—a substance endowed with *inertia*, and capable, in accordance with the established laws of motion, of imparting its thrill to other substances. If there is one conception more firmly fixed in modern scientific thought than another, it is that heat is a mode of motion. Ask yourselves how the vast amount of mechanical energy actually transmitted in the form of heat reaches the earth from the sun. *Matter* must be its vehicle, and the matter is according to theory the luminiferous ether.

Thomas Young never saw with his eyes the waves of sound; but he had the force of imagination to picture them and the intellect to investigate them. And he rose from the investigation of the unseen waves of air to that of the unseen waves of ether; his belief in the one being little, if at all, inferior to his belief in the other. One expression of his will illustrate the perfect definiteness of his ideas. To account for the aberration of light he thought it necessary to assume that the ether which encompasses the earth does not partake of the motion of our planet through space. His words are:—'The ether passes through the solid mass of the earth as the wind passes through a grove of trees.' This bold assumption has been shown to be unnecessary by Prof. Stokes, who proves that, by ascribing to the ether properties analogous to those of an elastic solid, aberration would be accounted for, without supposing the earth to be thus permeable. Stokes believes in the ether as firmly as Young did.

I may add, that one of the most refined experimenters in France, M. Fizeau, who is also a member of the Institute, undertook to *determine*, some years ago, whether a moving body drags the ether along with it in its motion. His conclusion is that *part of the ether adheres* to the molecules of the body, and is transferred along with them. This conclusion may or may not be correct; but the mere fact that such experiments were undertaken by such a man illustrates the distinctness with which this idea of an ether is held by the most eminent scientific workers of the age.

But while I have endeavoured to place before you with the utmost possible clearness the basis of the undulatory theory, do I therefore wish to close your eyes against any evidence that may arise of its incorrectness? Far from it. You may say, and justly say, that a hundred years ago another theory was held by the most eminent men, and that, as the theory then held had to yield, the undulatory theory may have to yield also. This is perfectly logical. Just in the same way, a person in the time of Newton, or even in our own time, might reason thus: The great Ptolemy, and numbers of great men after him, believed that the earth was the centre of the solar system. Ptolemy's theory had to give way, and the theory of gravitation may, in its turn, have to give way also. This is just as logical as the former argument. The strength of the theory of gravitation rests on its competence to account for all the phenomena of the solar system; and how strong that theory is will be understood by those who have heard in this room Professor Grant's lucid account of all that it explains. On a precisely similar basis rests the undulatory theory of light; only that the phenomena which it explains are far more varied and complex than the phenomena of gravitation. You regard, and justly so, the discovery of Neptune as a triumph of theory. Guided by it, Adams and Leverrier calculated the position of a planetary mass competent to produce the disturbances of Uranus. Leverrier communicated the result of his calculation to Galle of Berlin; and that same night Galle pointed the telescope of the Berlin Observatory to the portion of the heavens indicated by Leverrier, and found there a planet 36,000 miles in diameter.

It so happens that the undulatory theory has also its Neptune. Fresnel had determined the mathematical expression for the wave-surface in crystals possessing two optic axes; but he did not appear to have an idea of any refraction in such crystals other than double refraction. While the subject was in this condition the late Sir William Hamilton, of Dublin, a profound mathematician, took it up, and proved the theory to lead to the conclusion that at four special points of the wave-surface the ray was divided not in *two* parts, but into an *infinite number of parts*; forming at those points a continuous *conical envelope* instead of two images. No human eye had ever seen this envelope when Sir William Hamilton inferred its existence. If the theory of gravitation be true, said Leverrier, in effect, to Dr. Galle, a planet ought to be there: if the theory of undulation be true, said Sir William

Hamilton to Dr. Lloyd, my luminous envelope ought to be there. Lloyd took a crystal of Arragonite, and following with the most scrupulous exactness the indications of theory, discovered the envelope which had previously been an idea in the mind of the mathematician. Whatever may be the strength which the theory of gravitation derives from the discovery of Neptune, it is matched by the strength which the undulatory theory derives from the discovery of *conical refraction.*

NOTE.

I would strongly recommend for perusal the essay on Light, published in Sir John Herschel's 'Familiar Lectures on Scientific Subjects.'

J. T.

LONDON: PRINTED BY
SPOTTISWOODE AND CO., NEW-STREET SQUARE
AND PARLIAMENT STREET

WORKS

BY

JOHN TYNDALL, LL.D. F.R.S.

Professor of Natural Philosophy in the Royal
Institution of Great Britain.

Researches on Diamagnetism and Magne-Crystallic Action;

Including the Question of Diamagnetic Polarity. With numerous Illustrations. 8vo. price 14s.

On Radiation;

The Rede Lectures delivered before the University of Cambridge, May, 1865. Crown 8vo. with Diagram, price 2s. 6d.

'Few men possess the remarkable faculty of making abstruse subjects connected with natural philosophy intelligible to ordinary untrained minds in the same high degree as the Author of this Lecture. It is an admirable exposition of the present state of our knowledge as regards radiation, and will be read with profit by all who desire to become acquainted with the subject.'
MEDICAL TIMES and GAZETTE.

Faraday as a Discoverer.

New and Cheaper Edition, with Two Portraits. Fcp. 8vo. price 3s. 6d.

'Professor TYNDALL'S Memoir of FARADAY as a discoverer is written in clear and vigorous English. FARADAY was a man of the loftiest aims, and was probably one of the greatest experimental philosophers the world has ever had. His character as a man of science, and the extent to which science is indebted to him, and the nature. method, and the precision of his discoveries—all these matters the reader will find well told in this volume.'
The LANCET.

'This welcome little volume contains three portraits—*Faraday the Philosopher, Faraday the Man, Faraday the Christian.* The portraits are drawn with a firm and clear hand, in a gentle and loving spirit, under the guidance of a deep insight. Men of science who clustered round FARADAY'S home in Albemarle Street will be pleased that the portrait of their distinguished chief has been trusted to the hands of one of the most eminent among themselves, whom FARADAY selected as his assistant and successor. The members of the much wider circle whose lives were illuminated by the rays of truth which beamed on them from that luminous fane of science where young and old, ignorant and skilled, were through so many years equally charmed, elevated, and instructed, will be grateful that the character, the labours, and the teachings of their master are herein transmitted to them by a fellow-pupil who neither in admiration nor affection falls short of their own. They will all give Professor TYNDALL'S work a profound welcome.'
MACMILLAN'S MAGAZINE.

London: LONGMANS and CO. Paternoster Row.

Sound ;

A Course of EIGHT LECTURES delivered at the Royal Institution of Great Britain. Second Edition, revised; with a Portrait of M. Chladni, and 169 Woodcut Illustrations. Crown 8vo. price 9s.

'Few scientific works have been more rapidly or more deservedly successful than this admirable treatise, which has been translated into French and German and republished in the United States. It is a beautifully written work, eloquent and poetical in style, correct and accurate in expression and thought. It cannot be praised too highly, or too widely diffused.' The STUDENT.

'The contents of Professor TYNDALL'S book are of so attractive a nature, and recommend themselves so strongly, not only to the dilettante lover of knowledge, but to those who are earnestly engaged in the cultivation of science, that we are not surprised a second edition has been speedily called for. Having already noticed at length, in our review of the first edition, the characteristic features of the work, the number and ingenuity of the experiments (in which Professor TYNDALL stands without a rival), the felicitous explanations and varied illustrations, we need here make no further remark than to say that the present is a reprint of the former edition with the exception of a chapter containing a summary of the recent researches of M. REGNAULT, written by himself. This contains some interesting observations on the propagation of sound in closed tubes, in which it is shown that the diameter of the tubes make a considerable difference in the intensity with which the wave is propagated through it, diminishing rapidly the smaller the section of the tube.....There are other interesting facts described in reference to the velocity of the propagation of waves, which we have not space to give, but which will well repay perusal.' The LANCET.

Heat a Mode of Motion.

Third Edition, with Alterations and Additions. Plate and 108 Woodcuts. Crown 8vo. price 10s. 6d.

'Beyond question the best and clearest popular exposition of the dynamical theory of heat that has yet been given to the public.' SPECTATOR.

'A want had long been felt among engineers for a clear and intelligible work on the mechanical theory of heat, and which should at the same time give an account of the scope of the experiments and numerical data upon which the theory is founded. This want is supplied by Professor TYNDALL'S book. The clear style of the work adapts it to the most ordinary capacities ; and the reader is raised to the level of these questions from a basis so elementary that a person possessing any imaginative faculty and power of concentration can easily follow the subject. Popular as Professor TYNDALL'S exposition is, we are convinced that the most accomplished man of science would rise from its perusal with an additional amount of information.' MECHANICS' MAGAZINE.

London: LONGMANS and CO. Paternoster Row.

CPSIA information can be obtained at www.ICGtesting.com
Printed in the USA
LVOW02s2142061013

355646LV00016B/927/P